SCIENCE PARKS

SCIENCE PARKS
DEVELOPMENT AND MANAGEMENT

Norma Carter

1989

Estates Gazette

THE ESTATES GAZETTE LIMITED
151 WARDOUR STREET, LONDON W1V 4BN

First published 1989
ISBN 0 7282 0130 5

This book was commissioned and produced for the Estates Gazette Limited by Leaf Coppin Publishing Ltd

Phototypeset by Input Typesetting Ltd, London
Printed by Hobbs the Printers Ltd, Southampton SO9 2UZ

CONTENTS

LIST OF PLATES

ACKNOWLEDGEMENTS

This book is based on discussions and information given by a large number of people involved in the planning, development, management and use of science parks. They are too numerous to list individually but I should like to thank them all: without their helpful and friendly attitude and their willingness to find time in their busy schedules such detailed insight into the operation of science parks would not have been possible. Chapters 4 and 5, in particular, are based almost entirely on knowledge gained by the author in this way.

In addition, I should like to thank Brian Worrall for providing invaluable data from surveys carried out by UKSPA. Ray Oakey has kindly given permission for the use of Figure 8, Chapter 6.

Thanks go also to Mandy Nelson for her diligence, patience and good humour in turning messy manuscripts into accurate typescript.

1. INTRODUCTION

Science parks can now be seen as a type of employment environment and a property market distinct from other forms of industrial development. Their importance in the property world means that no practitioner can afford to ignore what is happening. There are signs that a significant resurgence of industry could be rooted in academia and that small businesses can be nurtured in the right environment. Evidence suggests that there is significant growth potential which offers considerable opportunities for developers and managers of industrial property. Science parks may be the most outstanding growth sector in the next couple of decades.

Science parks are not established primarily as property ventures, but to fulfil two main objectives: to encourage technology transfer by reducing the time lag between innovation and production of new products in a research and development environment; and to provide means of encouraging the establishment and growth of high technology industry in an area in order to generate wealth and employment. The property, although essential, is a means to an end. Land and buildings, however, are not enough in themselves to achieve these objectives.

The objectives of science parks have implications for the type of tenant, the location and the buildings. The tenants will be innovative and companies relatively small. Sites will be located close to academic or research centres and buildings often designed for multi-occupancy. The exceptionally volatile nature of the users, the novel locations and customised buildings will influence design, funding and management.

Science parks are a new type of property investment with a different degree of risk from other industrial property. They share many of the characteristics of risk capital investment and are more akin to that than to conventional institutional property investments. Although this may be an outstanding growth sector, in order to realise that opportunity new investment vehicles may well be required.

Industrial Property

Property for industrial use has changed considerably during the 1980s – both the buildings themselves and the spaces in between. Conscious attention has been given to image and to its marketing. One manifestation of this is the use of the word 'park', to suggest pleasant, comfortable, even desirable, surroundings. Other such words, used to imply the forefront of producing new products, an entrepreneurial environment, success and a new age in British industry, are 'science', 'technology', 'high-tech' and 'business'. Unfortunately for the tenant and the investor, the reality is frequently different. The premises, their management and their neighbours may not be what the advertising brochures led one to expect. Terms such as 'research park', 'science park', 'innovation centre', 'technology centre', 'technology exchange', 'technology park' and 'business park' have been used more often to create image than with detailed thought as to the eventual occupier.

The users are the most important ingredient in any industrial property. They ultimately set the tone of a development. Premises need to be suitable for the users' needs in terms of space, services, facilities, costs and management. An under-occupied building is clearly not the objective of any developer, thus it is important to do detailed market research so that location, siting, premises and management are right for intended users.

Industrial users have changed: decline in the older heavier industries and growth in microelectronics, biotechnology and pharmaceuticals; decline in the number of manufacturing jobs and an increase in service sector employment; decline in the north and inner cities and growth in the south and small towns. These are the general trends. There has also been a tendency for all industrial concerns to become more sophisticated in terms of both production processes and products in order to combat higher costs, especially labour, and to compete better in the marketplace. The spread of process technologies to countries outside Europe and North America has led to increasing concentration on product and systems innovation. Product design is seen as the main determinant of both cost and marketability. Research, development and design are thus increasingly important activities in UK industry.

These new industries are footloose in a way which traditional ones, with their static location factors, were not. This does not mean location factors are not important, indeed they are. Rather it implies fiercer competition between regions and nations to attract these growth industries.

The spatial requirements of high technology users are also different from traditional industrial users. There is a need for a mixture of industrial activities under one roof (manufacturing, assembly, administration, customer services, repairs, sales, dispatch, showroom, storage, training and research and development), both because the activities are more integrated than in traditional industrial processes and because there may be operational reasons to change the balance between the activities at short notice.

In the UK there has been rapid growth in the newer industries in certain areas of the country where the demand for new property from industry and for better working conditions from the well-paid work-force have combined to produce a new type of industrial property which is modern in appearance, well provided with car parking and pleasant surroundings, but which is also adaptable internally to a variety of uses. High technology industry has thus been perceived by developers as a new kind of user which is fairly homogeneous and which, because it is a growth sector, can afford high rents. Furthermore, it has been assumed that because of a concern for its image, high technology industry will pay for high specification premises and landscaped environments. The conclusion has been that a type of 'high-tech' property can be offered which is adaptable in use, is capable of providing for a high percentage of office space and has the right modern image. The term 'high-tech' in this context refers to the appearance of the building rather than the user's activity which may in fact be traditional light industry, warehousing or offices.

Uncertainty (and frequently ignorance) about both the newly emerging high technology industries and the changes in industry generally have meant that developers and funding institutions have wanted to minimise the risks by constructing buildings suitable for a variety of uses. Thus, internal flexibility to adapt to different uses and users has been important from not only the user's point of view but also the developer's. In addition, the phenomenal growth in some areas (the Heathrow triangle, M4 and M3 corridors, Cambridge and the M25 belt) compared with the depressed industrial markets elsewhere in the country has persuaded the financial institutions as well as property development companies to become involved in this sector. An established market now exists for this 'high-tech' property. The designs are generally based on the assumption that users are flexible and tenants with good covenant can be obtained on standard institutional lease terms. The successful letting of individual buildings may well be fortuitous. In large-scale developments or those phased over a period of years, success is more likely to be achieved if designs are based on more detailed analysis of user requirements.

The name, it could be argued, does not matter so long as the development is profitable. Unfortunately, a lack of clear-cut definitions and of consensus as to what particular terms mean can lead to serious misunderstanding between owners, developers, funding institutions, surveyors and planners. If the purpose of the property industry is, like any other business, to provide the customer with what he or she wants, at the right time, in the right place and at a price the customer is willing to pay, then terminology is important. Creating the right product for the customer, i.e. the user, will become more important as the new technologies and individual companies grow and as companies and employees become more critical of the premises which they rent or in which they work.

Collaboration between Industry and Academia

Collaboration between industry and universities has been encouraged by successive UK governments. During the mid-sixties a variety of policies were directed towards ensuring that the national economy benefited from evolving technologies and new scientific discoveries. Initiatives included ten new technological universities and support for the establishment of industrial liaison officers in universities. The latter was part of a fund set aside in 1967 by the University Grants Committee to give short-term assistance for specific projects which would aid university–industry collaboration. Each project was expected to become self-financing in a year or two and to be capable of being absorbed into regular university commitments. Courses orientated towards industry were also introduced. In the same year, a joint committee of university vice-chancellors and senior industrialists was set up to study the relationships between universities and industry in the field of research. This recommended industrialists to take a greater interest in university activities, including staff secondments, student placements, contract research and use of university consultants.

Since then numerous schemes and programmes have been developed. The general aims of these have been to raise the level of technological capability in industry and to ensure that knowledge gained through academic research is transferred to industrial use. The Collaborative Awards in Science and Engineering (CASE) scheme, the Teaching Company Scheme and the Alvey Programme have been important in this context.

Since 1981, the Government's desire to seek ways of containing public expenditure and to encourage public bodies to attract more funding from the private sector has had significant effects on the operation of universities and on research. In particular, priorities have been reassessed and emphasis given to the commercialisation of activities.

Expenditure on civil research and development in the UK compares unfavourably with the USA, West Germany, France and Japan: whether comparing total expenditure, growth in expenditure, or the contribution from government or industry.[1] On the other hand, government commitment to defence research and development spending is much higher than in all the main industrial economies except the USA.[2] This has relevance not only for fields of research pursued but also means a Ministry of Defence monopoly on the sales of many high technology companies: a factor of significance to many science park companies.

Government support for innovation, however, has increased since 1979 especially in the small firms sector. It is part of national and regional economic policies and is undergoing change which is likely to benefit science park companies. The Enterprise Initiative launched by the Department of Trade and Industry in January 1988[3] aims to give support for initiative in a more targeted way. There will be emphasis on small firms, technological innovation and inner city regeneration. The Research and Technology Initiative will focus on collaborative projects between firms and universities. The Business Education Initiative includes the regional technology centres and continues the Teaching Company Scheme.

Science parks are just one of the initiatives which endeavour to encourage technology transfer. It is important to see them in the broader context of other initiatives directed towards similar objectives.

References

1. Monck, C. S. P., et al., *Science Parks and the Growth of High Technology Firms*, Croom Helm, 1988, pp. 23–38, reviews research on the relationship between technical progress and economic development.
2. Advisory Board for the Research Councils, *Report of the Working Party on the Private Sector Funding of Scientific Research*, HMSO, May 1986, pp. 5–6.
3. Department of Trade and Industry, *DTI – the Department for Enterprise*, Cmnd 278, HMSO, January 1988.

2. EMPLOYMENT ENVIRONMENTS

The basic purposes of providing new planned employment environments remain the same as those of the industrial estates, trading estates and office complexes of the past: to provide suitable accommodation for industry and business together with development and investment opportunities. Science parks are one of the more recent forms to have emerged. Their property requirements will be better understood by an appreciation of their origin and growth and their place within the range of employment environments.

Origin of Science Parks

United States of America

The science park concept originated in the USA and took shape during the 1950s,[1] the first example being the Stanford Industrial Park (Silicon Valley, California). It was an initiative associated particularly with the Dean of Engineering who, in the late 1940s, helped two students, Hewlett and Packard. It led in 1950 to 660 acres of the 8,100-acre site being set aside for science-based industrial development. By 1980 the park had eighty occupants and 26,000 employees including the giant Hewlett Packard company. Research Triangle Park, North Carolina, another much-quoted example, began in 1959 with the commitment of three prestigious universities and a hundred and sixty acres of cheap land. Although development was slow in the first few years, it is now the largest planned research centre in the USA, occupying some 5,500 acres and employing 22,000 people. Only limited manufacture is permitted on the park, thus some companies have brought extensive manufacturing capacity into the neighbouring region. A third well-known park is the MIT Technology Square, Massachusetts, founded in 1961, and the related complex of industrial development in the surrounding area along Route 128. This was attracted by the wealth of expertise in electronics in MIT and has generated a large number of jobs.

Science parks in the USA are very much larger than those in the UK. There are, however, some features which provide useful lessons for those involved in science park development in the UK. First, many science parks, including some attached to major universities, have failed. Secondly, many science parks, including those cited, have had the advantages of cheap land and an aggressive university able to attract and hold top scientists and technologists. Thirdly, science parks have often been initiated by, or in, states suffering unemployment and a depressed economy (e.g. Research Triangle Park is in an area which was suffering from the effects of a run-down tobacco industry). Fourthly, seed capital and venture capital has been available on favourable terms (see below, p. 51). Finally, there is a need for regional as well as site infrastructure if spin-off benefits and the multiplier effect are to be gained.

Western Europe

Science parks in other western European countries also make interesting comparisons. Although the UK is some twenty years behind the USA it has been in the forefront of science park development in Europe. A comparative study[2] of West Germany, France, Belgium, the Netherlands and the UK shows that in 1980, when the UK had three science parks (Cambridge, Heriot-Watt and Birchwood[3]) with seventy-six companies, the only other countries to have any were France and Belgium. In France in particular the parks were much larger (three parks and 275 companies), and the emphasis was on attracting larger, already established research-based companies. By the end of 1985, when the UK had twenty-one science parks (excluding Birchwood and other 'commercial science parks'), accommodating 288 companies, West Germany was the only country to rival the UK growth (eighteen parks and 270 companies). In their emphasis on smaller companies, innovation and technology transfer, West Germany and the UK are similar. Other countries have been slower to develop innovation centre initiatives.

Some useful lessons can be drawn from these European comparisons. Firstly, all countries rely heavily on government (central and local) funds for infrastructure and buildings. In the UK, however, the contribution is the lowest. Secondly, the most successful parks are those with a strong relationship between science park management and the university (or other higher education institution), with the latter involved in that management. Thirdly, science parks play an important role in stimulating economic growth but an essential prerequisite is the intellectual catchment area provided by the academic institution.

United Kingdom

The Cambridge Science Park The most successful of the first generation science parks in the UK, it can usefully be compared with the US examples. It was first proposed at the end of 1969 by Trinity College, a university college rich in both land and scientific knowledge. There were three important stimuli to its establishment.[4] The first was interchange between science-based industry and the university. Linkage was seen as a two-way process. Leading academics felt applied research to be more meaningful than pure research. They wanted feedback from industry and placements for graduate students. As well as the internal pressures there were important external factors. The Wilson Government was keen to see evidence of the 'white heat' of technology in Great Britain. Cambridge felt threatened by the new technological universities to which resources were being diverted. Also, there was much discussion of the 'brain drain'.

Secondly, there would be greater flexibility of employment opportunities. Industry was being forced out of Cambridge by restrictive planning policies and the city was dominated by the university at the expense of other land users and employers. The university, industry and many politicians wanted to see increased opportunities for industrial growth.

Thirdly, there was the chance of development profit. The Senior Bursar of Trinity College, responsible for the management of its large property portfolio, had the vision (and persistence) to see long-term opportunities for a parcel of land which was mostly derelict. It is interesting to note that the academics who were involved in setting up the Cambridge Science Park were much impressed with US examples, especially Route 128.

The Cambridge Science Park developed very slowly and cautiously until 1979 when Napp Laboratories committed themselves to a ground lease of twenty-eight acres to accommodate their consolidated company and headquarters. It was the first large company to come on to the park, and by occupying the whole of Phase 2, it at once doubled the area of development. Also, by taking a ground lease at an initial premium it provided substantial funds for the college to open up the next phase for development. Napp had chosen a prominent site (Plate 1) thus leaving undeveloped land between it and the almost completed Phase 1. Phase 3 was then able to benefit from contributions to the highway and other infrastructure works made by the Napp development. The opening of the Napp building in September 1983 was a turning point in the development of the Cambridge Science Park.

More recent UK science parks which began to emerge a decade after Cambridge can point to some remarkably similar factors related to university–industry linkages, employment needs and development opportunities. Figure 1 summarises the main factors.

The University of Warwick Science Park This shows how local authorities as well as universities may play a positive part. As the effects of the economic decline of key industries in Coventry and others parts of the West Midlands became apparent in the mid- to late seventies, Coventry City Council was increasingly conscious of the scientific resources available within the university and polytechnic and their lack in the local manufacturing sector. During this time a local plan was being formulated for the south-west area of Coventry which includes the university. The important issue of the future of the university and the surrounding land

Fig. 1 Factors contributing to the emergence of science parks

1.	Industrial decline and consequent demand for job and wealth creation	→	Need to nurture new businesses in the growth industries
2.	Perception that universities should be seen to be more relevant to industry	→	Need to encourage technological transfer
3.	Demand from universities for industrial linkage for staff research, student placements and premises for spin-out companies	→	Need for appropriate large industrial companies and small premises to be nearby
4.	Pressure on university finances, both actual cuts and exhortations to be more 'commercial'	→	Use of university land for income producing purposes
5.	Industry's need to improve the quality of its products and the efficiency of its processes	→	Use of university equipment and scientific and technological knowledge

(owned by the city council and used for sports facilities) was included in the draft South-West Area Action Plan in 1979. Investigations on science park development were carried out by the city council the same year. In 1981 it initiated discussions with the university.

The university which in its 1964 development plan had anticipated accommodating 20,000 students by 1995 was having to reassess its role and priorities owing to government cut-backs in university expansion. Changes in the university were thus a significant factor in the establishment of the science park. The importance of a science park as an agent for industrial regeneration in the West Midlands was recognised at the meetings between the university and the city council. The proposal was approved in principle in July 1981. Two matters which were stressed are of interest: the need for strong university leadership and the need for development finance. The university approached the former West Midlands County Council which became strongly supportive after the 1981 county council elections and the setting up of the West Midlands Enterprise Board. Subsequently, Warwickshire County Council was approached and agreed to become involved.

Accommodation for High Technology Firms

Property requirements will depend on several factors, the most important of which are the industrial sector within which the user operates, the stage in the firm's own growth and its organisational structure.

High technology industries are diverse in character. Technological change in industry takes place through either product or process innovation. Innovation may be incremental and continuous or radical and discontinuous. High technology industries are those concerned with the design and production of technologically advanced products or those exploiting high technology procedures in their production or warehousing systems. Various attempts have been made[5] to categorise those industrial sectors which are technologically sophisticated using measures of product innovation per employee, extent of use of microelectronics in products or processes, expenditure on research and development per employee and proportion of professional, technical and scientific personnel in the work-force. Instrument engineering, electrical engineering, mechanical engineering, vehicles, coal and petroleum products, chemicals and allied industries and professional and scientific services are the Industrial Orders (Standard Industrial Classification, 1968) from which high technology firms can be expected. Figure 6, Chapter 6, shows the specific industrial sectors of existing firms on UK science parks.

The leading edge of high technology industry[6] is where new technologies are developed, and where knowledge is translated from research into industrial production. Hence the use of the terms 'leading-edge industry', 'science-based industry' and 'knowledge-based industry' to distinguish innovative industry from high technology industry in general. The latter is primarily concerned with the exploitation of existing technologies for the production of goods and services in the areas of precision engineering, electronics, pharmaceuticals and biotechnology. High technology in general includes mass production and large office complexes but very little research, either in-house or contract. Evidence shows that there is a clear link between internal research and development effort and the prosperity and growth of a firm.[7] It is also apparent that the most effective innovations may result from situations where the strengths of small and large firms are combined. Small firms can react more quickly, are more willing to take risks and have better internal communications than large firms. The latter, however, have better access to resources, markets and external networks of contacts.[8] Thus small firms often play a crucial role in the diffusion of new technologies.

The stage in its growth is of great importance in determining a firm's property requirements in terms of both actual accommodation requirements and ability to pay. The growth cycle can be better understood by examining two interrelated processes: the innovation–production cycle (see below, pp. 67–8) and the life cycle. Although they are not models of the behaviour of every firm and their interrelation will vary in individual cases, they do provide a valuable insight into the way firms may change and the demands of particular tenants.

Life-cycle theory suggests that there is a relationship between the maturity of a firm and its accommodation requirements. In the early stages, premises must be cheap and flexible both in physical and lease terms and be conveniently located with regard to both previous place of employment and residence of the key staff. In the intermediate stage, the firm's products or services become known and it, rather than the personnel, is recognised as a supplier of excellence. This is a stage of rapid growth when image is becoming important. A mature firm shows slower growth and projection of a corporate image is extremely important. This will include a prominent site and a distinctive image portraying innovation and success but also a sense of security to both customers and staff. Spin-out firms and mergers may occur at any stage but especially from the intermediate stage onwards: the cycle thus recommences or is cut short.

Organisational structure of a firm as well as historic reasons related to the way in which it has evolved will affect its spatial arrangements. Some are concentrated locationally so that all activities take place on the same site. Other firms may have several sites which all engage in similar activities, perhaps each making a different range of products. Many high technology firms, however, carry out different activities on different sites. Thus, the actual activity taking place on a particular site may be routine manufacture, warehousing or distribution, even though the firm as a whole does carry out significant research and development and is at the leading edge. Each activity will have different location and premises requirements. Research and development will require proximity to related research and industrial activities and highly qualified staff who in turn will demand a good living environment. Manufacture will require a pool of reliable, appropriately skilled labour as well as access to supplies and markets: many high-tech products require only low-tech mass production. Distribution and warehousing will require a good position in relation to national, even international, transportation networks. Some will require custom-designed building, others will have more universal requirements. Some will be image conscious, others not. All will be cost conscious.

New Employment Environments

Terminology

The terms 'science park', 'business park' and 'technology park' are often used interchangeably by the development industry. They are used by local authorities keen to attract industry and employment opportunities, by academic institutions keen to improve their credibility and their income, as well as by developers and their agents to imply an up-market image and thus to help achieve above-average rents. In none of these cases is there much regard for the occupier: they are all marketing ploys. Despite frequent misuse, the meaning of these terms is important (see above pp. 1–3), although there are problems in setting down rigid definitions either because these will be too exclusive or too inclusive or because they fail to take account of the changing nature of employment environments. A scheme is as much defined by its tenants as by its physical characteristics.

Classification of Employment Environments

A more useful approach to both understanding and using the terminology for new employment environments is to envisage the whole range of planned employment estates as a continuum from research centres at one end through science parks and business parks to conventional industrial or trading estates at the other. Each scheme or completed development can then be placed in the continuum according to its particular characteristics. This approach to classification will enable a better appreciation of the nature of the development whilst avoiding the pitfalls of over-rigid definitions.

Most important when classifying an individual development in terms of employment environments are the characteristics of the user and the objectives and resources of the developer. The interrelationship between these will largely determine the development which results.

The developer's character and the resources available to it will have a strong influence on its objectives. A pension fund, development company, university, local authority or central government agency will have its own reasons for becoming involved in development. Each will have access to different resources of finance, land and expertise which will be used to further its objectives. These in turn will affect the purposes of any particular development.

A university establishing a research park to attract prestigious research organisations and provide space for its own research activity will produce a very different development from a property company looking for a healthy developer's profit by catering for footloose high technology firms. Both will be different from an innovation centre set up by a local authority in the centre of a conurbation which is anxious to attract firms with growth potential. The relative importance attached to non-property objectives, such as encouraging technology transfer and providing for new jobs and the management arrangements adopted, will be critical factors in determining the yield sought and achieved. All will affect the nature of the development.

Science Parks

A science park is defined by its objectives, its occupants and its management. The non-property objectives of technology transfer and wealth and job creation by encouraging the establishment and growth of knowledge-based industry are of primary importance. They are concerned with creating confidence in a locality as well as a focus for knowledge-based industry linked to centres of excellence in research. More specific objectives may include provision for the following:

1. a bridgehead for the transfer of knowledge between academics and industrialists;
2. a mechanism whereby companies can spin out from universities;
3. a means by which academics can keep in touch with commercial realities;
4. a culture of interchange and collaboration between tenants;
5. involvement in the regeneration of the cutting edge of British and European technology.

These objectives are more important than development or investment yields. Nevertheless, it is important to satisfy both user and investor: the costs of land, buildings and maintenance cannot be ignored. Furthermore, since science parks are concerned with creating demand from tenants, the physical as well as the business environment is part of the image marketed to stimulate that demand.

It is also worth noting that universities themselves differ in terms of financial and land resources, expertise, location and attitudes: some are more property-conscious than others. Although, historically, universities have been the main institutions linked with science parks, other higher education institutions are also involved. In addition, research institutions and industrial companies may have the capabilities and incentives to become involved (see below, p. 29).

A working definition of a science park is a planned property initiative which:

1. has formal operational links with a centre of research (which may be an academic or research institution or a large industrial company with its own substantial research facilities);
2. is designed to encourage the formation and growth of knowledge-based businesses; and
3. has a management function which is actively engaged in the transfer of technological and business skills.[9]

'Science park' is thus used as a broad term to include a range of property initiatives variously described as 'research parks', 'innovation centres', etc., where they meet the criteria set out above.

Most science parks provide a type of property which has little competition and which is relatively expensive compared with typical industrial sheds for that area. Thus, firms need to be fairly confident before moving into a science park unit, a fact which is inconsistent with fostering start-up businesses, since these require low occupational costs and easy-in/easy-out terms. The developers and managers of science parks must come to terms with the conflict inherent in satisfying both property and non-property objectives. In this context the range of activities and quality of management support will be especially important.

References

1. This section is based on unpublished reports from various British university academics who have been on 'fact-finding tours' of US science parks as part of their own feasibility studies.
2. Sunman, H., 'The growth of science parks in Europe', UKSPA Conference, London, December 1986.
3. Birchwood Science Park, Warrington, which is included in ibid., is classified in the present study as a business and technology park rather than a science park. The initial ideas of linkage with academic institutions such as UMIST and Salford University have not materialised. The discussions which took place in the early 1970s were not followed through. As commercial and industrial companies moved to the Birchwood site, Warrington Development Corporation saw little point in developing linkages. Since the early 1980s the academic institutions themselves have developed their own initiatives adjacent to their campus. British Nuclear Fuels Limited (BNFL) which is now a major land user and employs some 4,000 people on the science park was well established before the science park idea was proposed. The presence of BNFL was not, however, a factor in setting up the Birchwood Science Park. Although BNFL has acted as a magnet to consultancy companies, it has not been a source of spin-outs nor does it offer support to new firms.
4. Carter, N., and Watts, C., *The Cambridge Science Park*, Planning and Development Case Study 4, Surveyors Publications, 1984, pp. 11–16.
5. For example: Oakey, R. P., Thwaites, A. T., and Nash, P. A., 'The regional distribution of innovative manufacturing establishments in Britain', *Regional Studies*, 14, 1980, pp. 235–53; Northcott, J., and Rogers, P., *Microelectronics in Industry, What's Happening in Britain*, Policy Studies Institute, London, 1982; Henneberry, J., 'Property for high technology industry', *Land Development Studies*, 1, 1984, pp. 145–68; Advisory Board for the Research Councils, *Report of the Working Party on the Private Sector Funding of Scientific Research*, HMSO, May 1986.
6. Department of Industry/Shell UK Ltd., *Helping Small Firms Start Up and Grow: Common Services and Technological Support*, HMSO, 1982, p. 42.
7. Oakey, R. P., 'Innovation and regional growth in small high technology firms', *Regional Studies*, 18, 1984, pp. 237–51.
8. Rothwell, R., 'The relationship between technical change and economic performance in mechanical engineering: some evidence', in Baker, M. J. (ed.), *Industrial Innovation: Technology, Policy, Diffusion,* Macmillan, 1979, pp. 36–59.

9. This definition is very similar but more open-ended than that used by UKSPA both to distinguish science parks from other industrial developments and as a set of criteria by which to determine eligibility of schemes for membership of UKSPA.

3. CHARACTERISTICS OF SCIENCE PARKS

Until 1982 the only science parks in the UK were the Cambridge Science Park and Heriot-Watt Research Park, both established in 1972. In recent years, however, the number has increased substantially (Fig. 1). At the end of 1987, thirty-three were in operation and a further five were under construction; over half of these parks are in assisted areas and well over a third in inner city locations. These figures exclude those developments which are called science parks but which have essentially conventional property objectives, such as Birchwood Science Park, Warrington (see p. 14, n. 3), but include developments using a variety of other names, for example research park and technopark (see above, p. 13, and Chapter 4, p. 36, Fig. 4).

Fig. 1 Number of science parks in the UK

Year (end)	*Science parks in operation*
1972	2
1982	3
1983	7
1984	16
1985	21
1986	28
1987	33

Growth

The first generation science parks were very slow to gain a momentum of their own. Figure 2 shows the build-up of tenants and floor space at the Cambridge Science Park. Some of the second generation parks developed during the early to mid-1980s, especially those with a proactive management, have shown much more rapid growth in their early years (Fig. 3, see pp. 18–19) and have been able to create a market in a relatively short time.

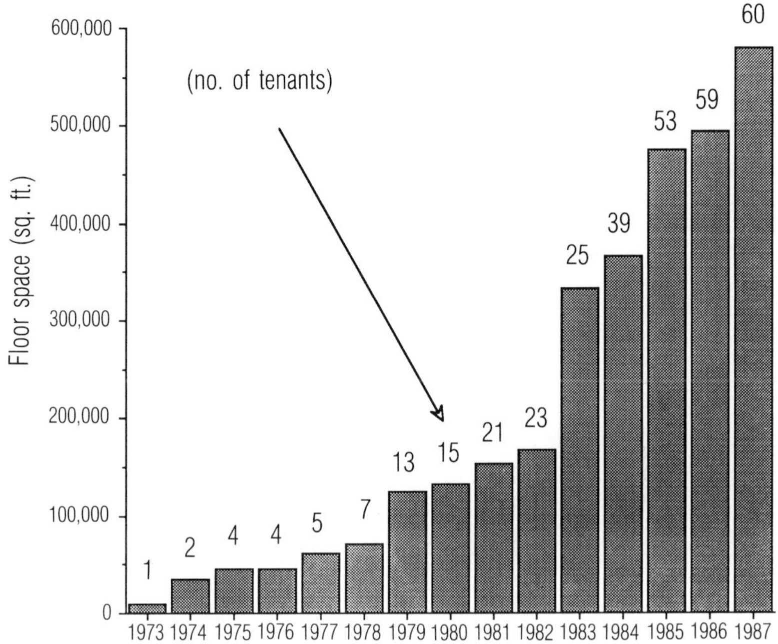

Fig. 2 Increase in floor space and tenants, Cambridge Science Park

There are 2,612,000 sq. ft. (242,655 m^2) (December 1987 figures) of completed floor space on UK science parks (Fig. 4, see p. 22). At the end of 1987, seven per cent was unoccupied, representing three months' supply at current rates.

The growth and success of science parks have far outweighed expectations. The pressures on higher education, local authorities, development agencies and large industrial companies outlined in Chapter 2, Figure 1 (see above, p. 9) have coincided with other factors related to new technology-based firms. First, their number has increased considerably in the past decade or so. Of particular importance has been the emergence of new technologies which lend themselves to exploitation by small firms, the encouragement given to independent business enterprise and the improved availability of venture capital (see below, pp. 51–2). Secondly, firms are demanding premises which are better than those on conventional industrial estates (see above, p. 1). Thirdly, some of them are consciously seeking to be part of a community of like-minded firms (see below, p. 72).

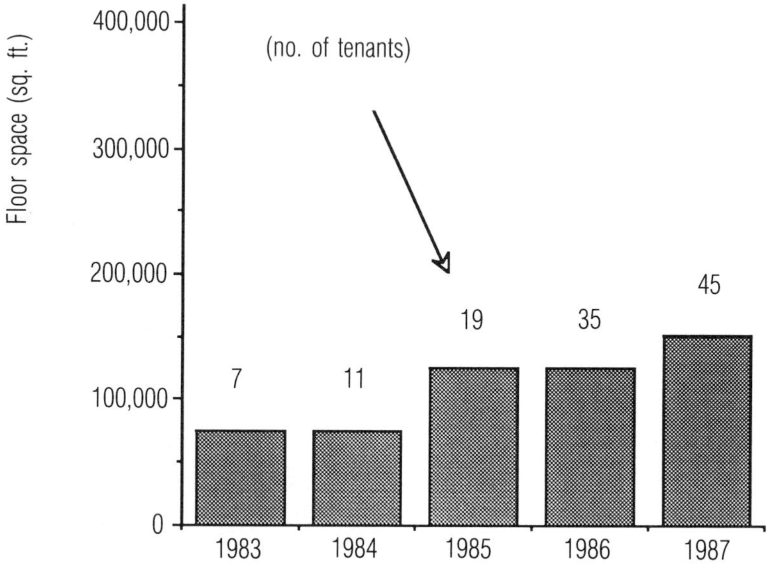

(1) Aston Science Park

Fig. 3 Increase in floor space and tenants

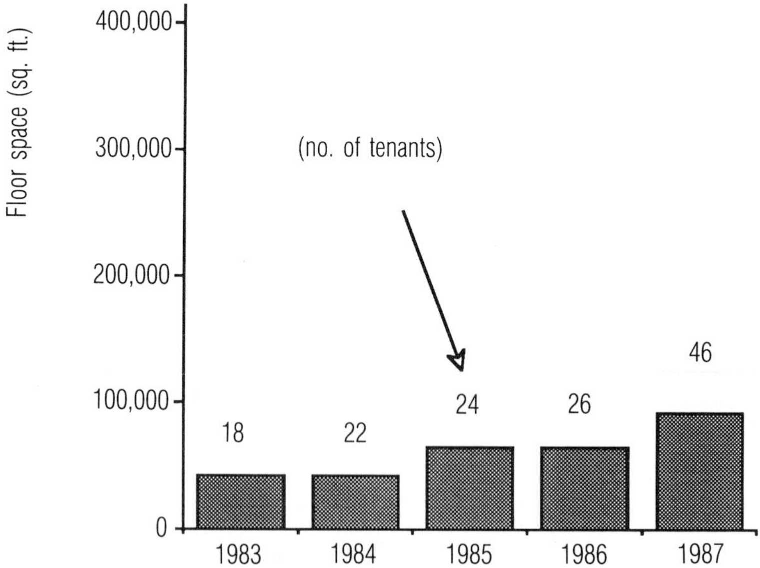

(2) Listerhills Science Park

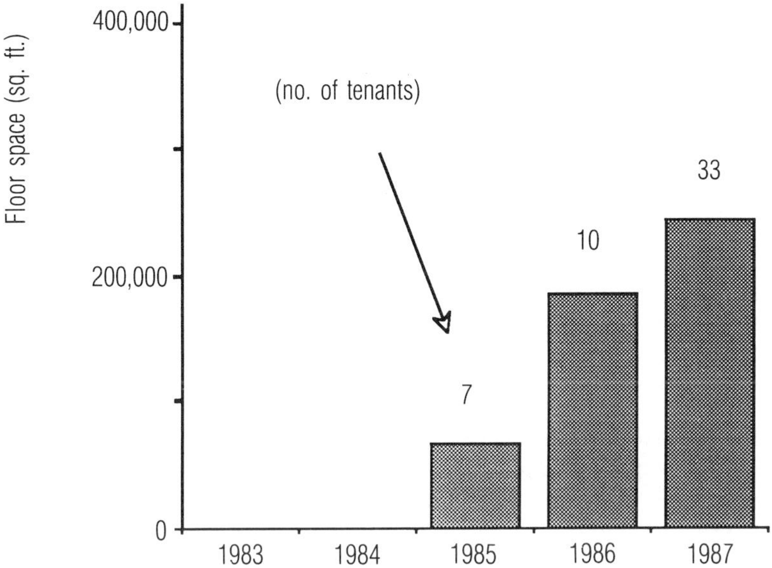

(3) Surrey Research Park

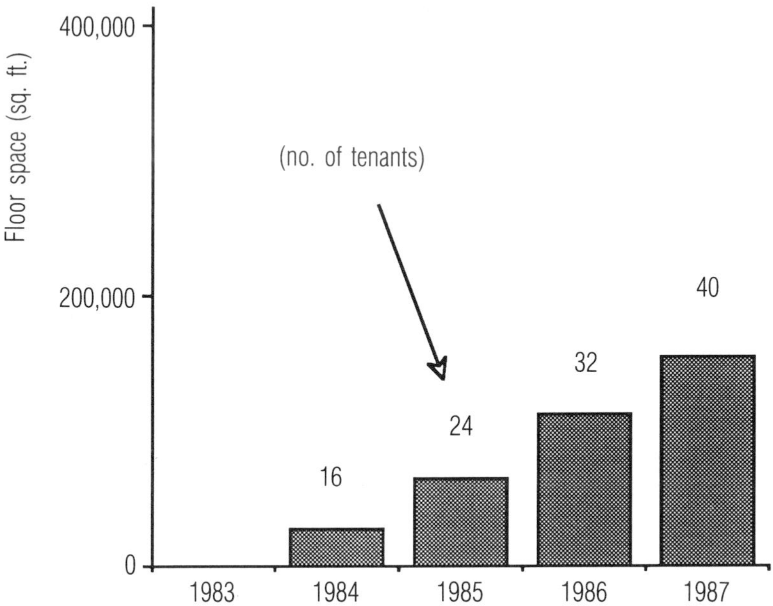

(4) Warwick Science Park

Location

Proximity to Academic or Research Institution

Science parks are mostly located close to universities or other higher education institutions (see above, pp. 13–14) since contacts are more likely to be established and continued if people are working in close proximity. Casual acquaintances are more likely to be followed up and new contacts sought when a minimum of time will be wasted in travelling. New and small companies do not normally like to set up meetings in advance since their daily schedules are so variable.

Proximity is thus especially important to small knowledge-based companies which need both technological and business support and whose growth is so dynamic. Although it may be argued that mature companies have little need to be close to a centre of academic learning, the most successful science parks do contain companies in all stages of development. Interaction between companies is at least as important as linkage with the university (see below, p. 72).

The question 'how close is close?' has been a matter of debate since the first proposals for a science park in Cambridge.[1] It was suggested that the site must be within cycling distance. Five miles was the radius of search area. The circumstances will be different for each town or city: cycling will not always be a consideration, but the facility for casual contact is very important. The Cambridge Science Park is only two and a half miles from the centre of Cambridge and Trinity College, but for the first decade, at least, that was felt to be far enough for start-up companies.

Around Birchwood Science Park, Warrington, opened in 1973, there is an exceptionally large concentration of prestigious academic institutions in Manchester, Salford and Liverpool, and a very good road network, yet no linkages materialised (see p. 14, n. 3). The Emersons Green proposals (see below, p. 34) include the three participating academic institutions having a physical presence on the science park site. In view of the distance to those institutions, this will be an essential prerequisite if linkage is to materialise.

Proximity will be influenced by the location of the higher education institution: inner city or 'green field', for example. It will also be affected by their land holdings as well as those of other development partners, especially local authorities, and by the opportunity costs of alternative uses. Some universities have significant land holdings, Cambridge, Surrey and Durham for example. Others have virtually no spare land and are sited close to the centre of large cities where land is costly.

Campus universities are generally more able to find an unused corner though inner city institutions may have under-used land nearby. There are indications that some institutions, may be prepared to enter into arrangements for the development of sites at a considerable distance from the campus (Imperial College, London, has undertaken negotiations for such a site).

Other Location Aspects

Access to other research and development contacts either in research institutions or large firms is also important. The more of such establishments in an area, the richer the potential network for contacts. These may be important for ideas, use of equipment, consultations or poaching staff.

Access to suitable labour, markets and supplies, advice and professional services including sources of funding for firms, to good residential and leisure environments and to national and international transportation networks are also important location factors for firms in new and competitive technologies.

Premises

In general, the appearance of science parks is of 'modern' buildings in a well-designed environment. The quality of materials and finish although variable is typically better than average for the area or than the alternative accommodation available. Trendiness is not sought, the more traditional appearance of brick and tile is preferred. Tenants at the Cambridge Science Park have indicated the importance of prestige and image offered by the site.[2] A survey at four English Estates science parks[3] showed the main factors influencing tenants' choice of location to be: prestige and image of the site (33 per cent); type of accommodation (26 per cent); car parking (24 per cent); availability of premises (17 per cent); and access to the higher education institution (15 per cent).

Premises should be designed to allow maximum internal flexibility to adapt to changing requirements: high sills, absence of internal load-bearing walls and good natural and artificial lighting. The ability to distribute cable services throughout a building is most important. Low energy costs and air-conditioning facilities may be required by some tenants. Generous car parking and security provision are essential. Entrances to both the site and individual premises are important: they should be visible and attractive, and finished to a high standard (Plate 2).

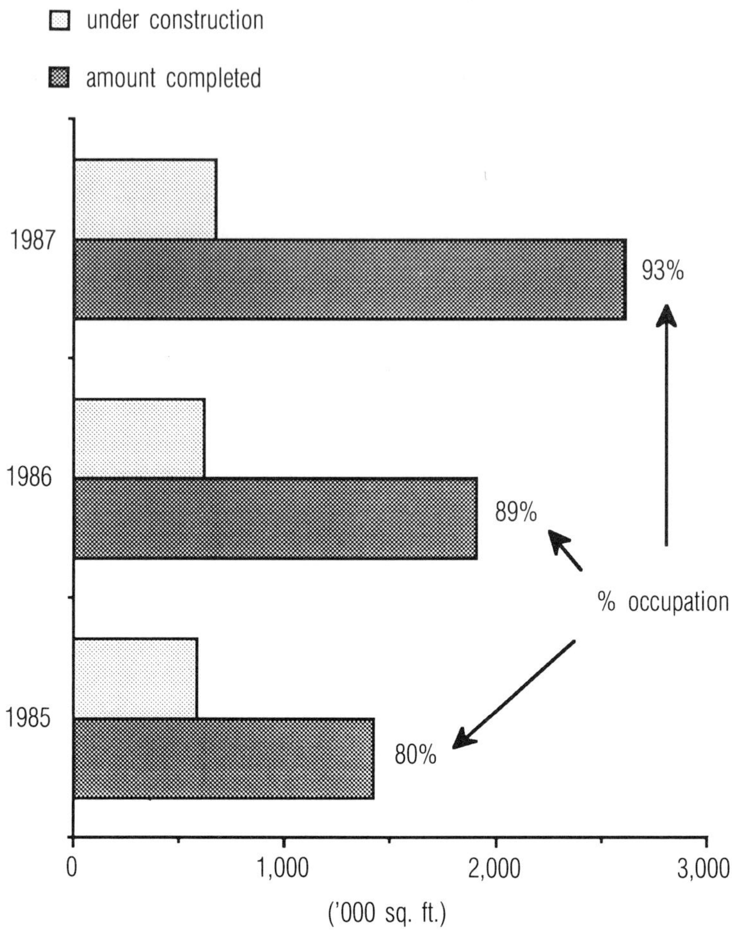

Fig. 4 Total floor space at UK science parks (*Source, UKSPA*)

Accommodation

Size and arrangement of units vary according to the anticipated market. Of the four identifiable categories of unit, only three are likely to be built speculatively.

Multi-occupancy buildings Designed as incubator units (Plate 3), these small separate units share entrance foyer and reception desk and usually a corridor and staircase. Thus communality is encouraged. Unit sizes range from 150-1,000 sq. ft. (14-93 m²). Some include rent-a-desk facilities. The buildings are frequently two storeys. The larger units may have a second access direct from the outside. The extent to which toilet and other facilities are shared will depend on the layout and size of units. Normally these units are finished and ready for immediate occupation. Lease terms are flexible, often outside the Landlord and Tenant Acts, and licences may be available. Service charges typically include heating. Instead of separate rent, service and rates charges, a single occupation charge may be made.

Small units with independent front doors These are usually in terrace or courtyard form with some parking in front of each unit (Plate 4) and additional parking elsewhere. They are designed both for firms able to operate on a day-to-day basis without external secretarial/receptionist support, and to be adaptable to changing requirements for the use of space, air-conditioning and other equipment. Gas, water, three-phase electricity and telephone connections are usually supplied. There is considerable variation in eaves height, type of access doors and standard of finish. Some are high enough for a mezzanine floor to be inserted and for substantial air-handling equipment to be installed within the roof space. Some have dual access, one through roller shutter doors (Plate 5). In general, such heights and doors are considered unnecessary, an internal floor-to-ceiling height of ten feet (3 m) meeting most needs. Units are designed on a modular basis, typically of around 1,200 and 2,400 sq. ft. (111 and 223 m²), so that more than one unit may be combined.

In some schemes attempts are made to combine these two forms of accommodation in a two-storey building. The smaller units are on the upper floor with communal access and the larger units on the ground floor, with independent external entrances (Plate 6).

Medium-size units Typically, these are between 5,000 and 10,000 sq. ft. (465-929 m²) and give the appearance of separate buildings, although in fact they may be linked (Plate 7). The design brief is likely to emphasise the flexibility to combine or split units as well as the need for internal adaptability. Tenants may have a landscaped area which is part of the demised premises; lease terms may be longer (twenty-five years) and be stricter full repairing and insuring leases.

Custom-designed units These are built for particular firms (Plate 8): either their activity requires an unusual building which would be difficult to let on the open market or the firm wishes to project its image through the style of its premises. Such buildings are normally at least 10,000 sq.

ft. (929 m^2) and possibly much larger. A ground lease is likely to be taken and either the firm will employ its own development team to carry out the work or the science park developers will put together a design-and-build package for the tenant.

Most UK science parks, to date, just offer accommodation in the first two categories (i.e. multi-occupancy and small independent units). Only a quarter of science parks have any medium-size accommodation. Figure 5 summarises the sizes of premises.

Fig. 5 Sizes of premises occupied by tenant firms in UK science parks (*Source, UKSPA*)

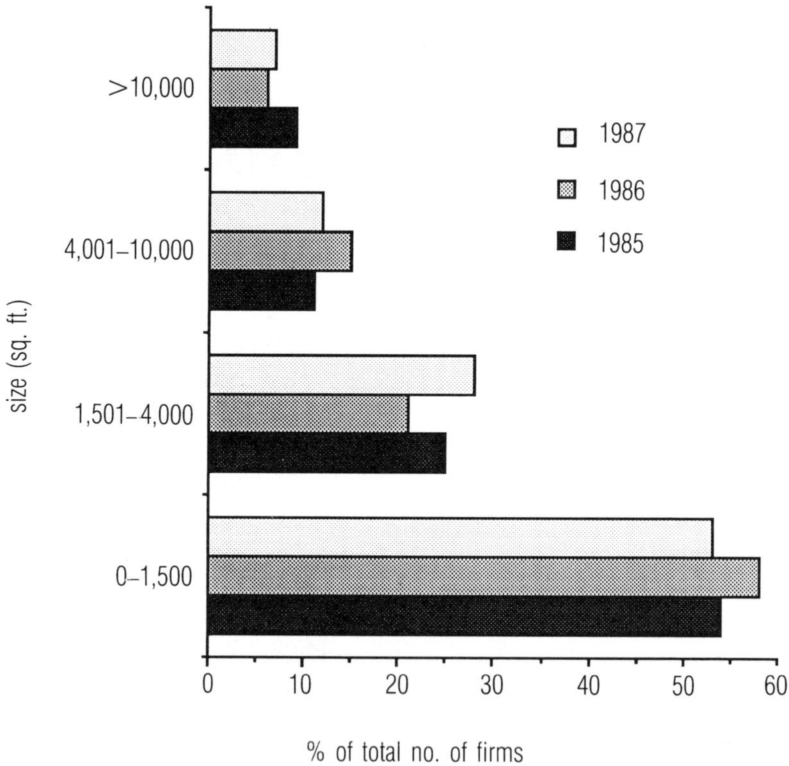

1 Napp building seen from entrance to the Cambridge Science Park, looking across Phase 3

2 Entrances to sites and premises should be visible and attractive
(1) Aston Science Park
(2) Mountjoy Research Centre

(3) St. John's Innovation Centre
(4) Wavertree Technology Park, small units

3 Multi-occupancy buildings designed as incubator units
(1) Surrey Research Park, Surrey Technology Centre
(2) Belasis Hall Technology Park, Belasis Court, Pavillion 1

4 Small units with independent front doors – at Bolton Technology Exchange

5 Small units with dual access, including roller shutter doors and service courtyard – at Highfields Science Park, Phase 1
(1) Front
(2) Rear

6 Combined incubator and small units
(1) Manchester Science Park, Enterprise House
(2) Mountjoy Research Centre

7 Medium-size units
(1) Cambridge Science Park, the first building, completed 1973
(2) Aston Science Park, Phase 2, Venture Units

8 Custom-designed units
(1) Cambridge Science Park, Cambridge Consultants
(2) Cambridge Science Park, Napp Laboratories

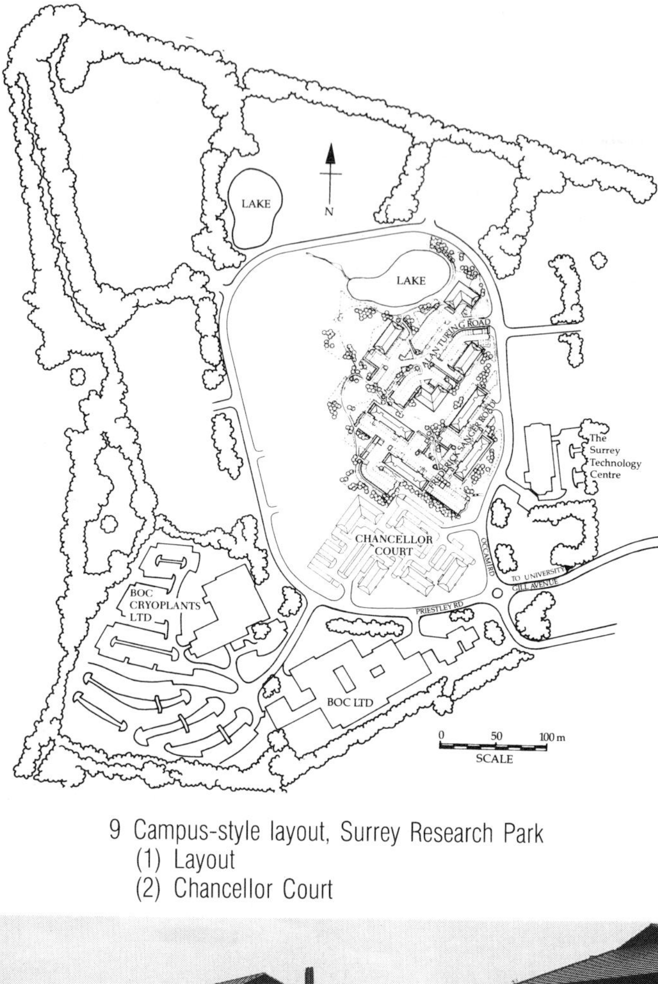

9 Campus-style layout, Surrey Research Park
(1) Layout
(2) Chancellor Court

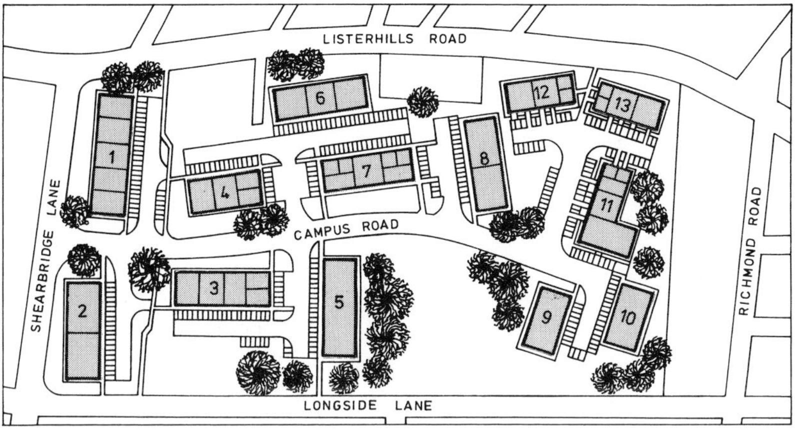

10 Urban-style layout, Listerhills Science Park
(1) Layout
(2) Typical unit

11 Refurbishment, Aston Science Park Phase 1
(1) General view of former metal-forging works
(2) Interior

Environment

Some science parks are low-density developments in a park-like environment; others, especially the innovation centres on small inner city sites, are essentially urban in appearance. Thus there is a considerable range of site densities and proportions of landscaped areas. Indeed, given the wide differences in overall size, location and type of user, average figures for site coverage are of little value. It is more useful to make comparisons with alternative schemes of similar type or in the same area and to look at the main categories of science park as they have emerged to date.

Campus-style The first two schemes, Cambridge and Heriot-Watt, are low density and well landscaped. At the Cambridge Science Park, where Phases 1, 3 and 4, which are fully developed, total 82.6 acres, the overall plot ratio is 0.20:1, and excluding the spine road and communal lake areas (24 per cent of the area), it is 0.30:1. Some later schemes, notably Warwick, Surrey, and Belasis Hall, continue that style (Plate 9). Warwick, when completed, will have an overall plot ratio of 0.22:1 (0.235:1 excluding distributor roads and communal landscaping), whilst on individual development sites the maximum plot ratio allowed is 0.25:1. Surrey aims to achieve 15 per cent overall site coverage: for individual land parcels the plot ratio is 0.25:1 and at least half the site area must be retained as open space. All these parks had the possibility of further land after the initial phase of development. In most schemes, the landscaped setting has been created by the developer. Strategic landscaping is thus an important part of the initial infrastructure and a critical factor in establishing the image. Belasis Hall, for example, has understood this: it was seen as essential to create a high-quality environment as quickly as possible by putting in strategic infrastructure, water features and planting for the first fifty acres, at the outset.

Urban-style Many of the 1980s science parks have been in inner cities, close to university departments and where land is tightly constrained. Although the external spaces are finished to a high standard and soft landscaping has been incorporated wherever possible, landscaped areas for communal use scarcely exist. The resulting building density is thus likely to be higher than in campus-style parks. Site coverage at Listerhills (Plate 10), which is made up entirely of single storey units, is 25 per cent. At South Bank, however, site coverage is 50 per cent and the units are in three storeys, i.e. a plot ratio of 1.5:1. Some of these developments began in refurbished premises. For example, Phase 1 of Aston Science Park, completed in 1983, consists of incubator units and related offices in a converted former metal-forging works (75,000 sq. ft./6,968 m^2 gross) (see Plate 11). Phase 1 of the Sheffield Science Park, the Cooper Building, is a former cutlery works which has been extensively refurbished to provide up to twenty-one work space units ranging in size from 170-1,480 sq. ft. (16-137 m^2).

In all schemes a high level of car parking provision will be required. Typically, one space to 250 sq. ft. (23 m^2) is provided, a standard which is already proving inadequate. Car parking must be conveniently and sympathetically sited, preferably landscaped and hidden from main approach roads. Attention to access and circulation for traffic and pedestrians, siting and design of storage areas, boundaries and signing (both for individual units and the site as a whole) are essential to a good environment. Given the car parking and landscaping requirements, the resultant site coverage will always be low, compared with conventional industrial estates, the footprint of buildings normally limited to 25 per cent. Maintenance of the common areas is a responsibility retained by the science park management.

Conference and Social Facilities

Some science parks have seen this as an important ingredient of successful development. As well as the more formal or organised requirement for conference and sports facilities, more casual social contacts are an important first step towards technical and business relationships. To date, few parks have been able to provide any facilities in the early stages of development. Those schemes adjacent to universities usually overcome the need for these facilities by arranging for science park employees to use those of the university, negotiating priority or favourable rates on their behalf. Such use of university facilities has the additional advantage of extending the network of contacts which may bring business advantages. Science parks at a greater distance from a university must solve the problem in another way. It was more than a decade before the Cambridge Science Park built a social centre. The Trinity Centre, which provides restaurant, bar and small conference facilities, opened in 1984 and the squash courts in 1986. All had been promised in the first design brief of 1971.[4] Innovation centres can sometimes provide small meeting rooms and a snack bar within the building. Some recent science park proposals include much more ambitious ancillary facilities. The problem remains: few developers are able or willing to meet the development costs.

Variety, as well as growth, is an important element in science parks. Most schemes have evolved to cater for an essentially local market of small high technology firms. The diversity shows that there is no standard formula for success. The linkage with the local economy rather than reliance on footloose research and development units of large companies indicates that the potential for science park growth is much greater than was once thought possible.

References

1. Carter, N., and Watts, C., *The Cambridge Science Park*, Planning and Development Case Study 4, Surveyors Publications, 1984, p. 15.
2. Tenant surveys carried out by the author in 1983, 1985 and 1987.
3. Southern, P., 'What are science parks? Variations in the product', UKSPA Conference, London, February 1988.
4. Carter, N., and Watts, C., op. cit., 1984, p. 31.

4. DEVELOPMENT OF SCIENCE PARKS

The first generation of science parks were speculative ventures by Trinity College, Cambridge, and Heriot-Watt University, Edinburgh, designed to increase industry–university liaison. They did not, in the early years, include speculative building. Few, if any, universities can afford to develop a fully fledged science park without external financial inputs. Given both the unfamiliarity of the science park market and the interests which both local government and development agencies have in stimulating the economy, the public sector has been an obvious source for development funding.

The second generation of science parks, opened since 1982, has been supply-led in that premises have been completed in advance of lettings. Usually at least one multi-occupancy building, sometimes a group of small units, is completed together with associated roads, car parking and landscaping. This form of development has made sense both in terms of nurturing new technologies and technology transfer and because many sites adjacent to urban universities were restricted in size. It also made sense because in the early 1980s it was difficult to attract medium or large growing companies into regions outside the south-east. Development funds, however, had to be committed in advance. Such commitment involved more risk and uncertainty than any investor applying normal commercial criteria could afford. Universities, local authorities and development agencies, however, can adopt longer term horizons and non-financial objectives and criteria more easily than the typical commercial development company or financial institution (Fig. 1).

Fig. 1 Total investment in land and buildings (*Source, UKSPA*)

UK science parks	*Investment in £m, to year end*		
	1985	1986	1987
Completed development	48	92	156
Earmarked for expansion	21	61	85
Total	69	153	241

The situation is changing. The market is more established but further huge capital investment is required. Local authorities and development agencies have used both internal resources and grant aid to finance development. These resources are, of course, limited and have in any case often been justified only as 'pump priming'. Thus there is a need for more private sector funding.

The Developers

Initiators

The initiative for setting up science parks has mostly come from universities or other higher education institutions. Those science parks where the university was the sole initiator are more likely to be in southern England, while the few science parks initiated by local authorities are all in the Midlands and north of England. In some cases it would appear that the idea originated simultaneously in the university and the local authority. Central government agencies, the Scottish and Welsh Development Agencies, and English Estates are the only other parties which have come forward to initiate any science park currently operational. The Belasis Hall development by ICI is the first to be initiated by the private sector. A number of other such proposals are emerging. There are also indications of a growing interest by government research institutes. The most advanced is a sixty-acre site adjacent to the Royal Signals and Radar Establishment (RSRE) at Malvern, a Ministry of Defence initiative which aims to increase the commercial benefit of resources devoted to defence research and development.

The role played by committed and determined individuals in influential positions in the initiation and continued support for the development of science parks is always significant. Frequently such individuals are powerful and influential personalities in a higher education institution; occasionally they have been important professional members in the local community with a particular vision.

Partners

Science parks generally require two or more partners who have been involved in some if not all aspects of development. A higher education institution has been concerned with all thirty-two science parks currently operational. Of these, only four have been linked to an institution other than a university and one of these was also connected to a university. Half the science parks involve one or more local authorities and about two-fifths one of the government development agencies. Only seven science parks have been developed by a university alone. In some cases, e.g. Cambridge and Surrey, this has been made possible by raising income from selling long leases. Figure 2 summarises the types of partners, though without taking into account their relative involvement. For instance, partners do not necessarily provide development finance. Indeed, in categories 3 to 7 of Figure 2 the higher education institution has usually contributed relatively little, if any, funding. Nevertheless, in most cases it has played a significant role in the formulation of ideas and of a development strategy including the assembly of resources. In some cases, it has provided the land.

Fig. 2 Partners in science park development in the UK

Partners	*No. of science parks (end 1987)* *Operational*	*Under Construction*
1. Higher education institution	3	–
2. Higher education institution Tenant companies	4	–
3. Higher education institution Local authority	11	–
4. Higher education institution Local authority Private sector	1	–
5. Higher education institution Government development agency	7	–
6. Higher education institution Government development agency Local authority	5	4
7. Higher education institution Private sector	2	–
8. Government development agency Private sector	–	1
Total no. of science parks	33	5

Investors

The public sector The public sector has provided 60 per cent of the development finance for UK science parks (Fig. 3). Universities have contributed development funds in several science parks, either from internal funds or by raising finance from the sale of ground leases (Cambridge, Surrey and Brunel). In other cases, university involvement has been restricted to a supportive role (e.g. Manchester and Nottingham).

Local authorities have provided substantial development funding, frequently covering the total costs of land and buildings. Some science parks (e.g. Warwick) have involved more than one local authority.

Where both local authorities and development agencies have played a role it has usually been the latter which has funded most of it, particularly in the assisted areas of the UK. Indeed, English Estates, already the

Fig. 3 Sources of development funding in UK science parks (*Source, UKSPA*)

Source	*% of total investment (end 1987)*
Universities	28
Local authorities	11
Government development agencies	21
Tenant companies	32
Private sector	8

largest developer and managers of industrial and commercial property, are also the largest developer of science parks. As the change of name from English Industrial Estates to English Estates indicates, their range of property types has widened to include offices, retail and leisure, as well as high technology premises. By their terms of reference, they are able to carry out development where private developers do not wish to commit themselves. Their involvement in science park development in assisted areas, where the main objective is job and wealth creation, may seem inevitable in retrospect, especially since 1981 when many of the more technological universities were badly affected by University Grants Committee cuts, but at the time it called for considerable vision, risk and courage on their part.

Listerhills Science Park at Bradford University was English Estates' first science park venture. Representatives of the university and city council together approached English Estates. The city had land adjacent to the university which was held for university expansion. Bradford University had had a severe cut in UGC funds in 1981: twenty-five per cent over four years. In the same year, local authorities were asked to dispose of unwanted land following the Local Government Planning and Land Act 1980. The Industry Act 1981 enabled English Estates to act in accordance with local markets but stressed that they should not compete with the private sector. English Estates were then looking for new ventures which would stimulate the local economy. Thus, the Bradford initiative came at an opportune moment. A feasibility study was commissioned and the decision was taken to proceed.

Feasibility studies were also carried out at Leeds and Hull. In 1988, English Estates (Yorkshire & Humberside) were participating in the development of four science parks, Bradford, Leeds and Hull, together with a project in Sheffield. English Estates (North West) have been involved in the Bolton Technology Exchange, the Wavertree Technology Park at Liverpool and latterly Salford University. English Estates (North) have developed the CADCAM Centre at Middlesbrough, the Mountjoy Research Centre at Durham and Belasis Hall Technology Park at Billingham. The extent of English Estates' involvement and whether active from inception (as at Listerhills and Belasis Hall) or as part of an existing venture (e.g. Sheffield and Salford) varies, but, overall, science parks have become an important flagship for English Estates, giving them prominence in the high technology market.

The private sector The private sector has, to date, had a smaller involvement than the public sector in funding science park development. Tenant companies themselves have played a significant role (Fig. 3). Where other private investment has been attracted the motivation has been largely philanthropic. In general, the financial institutions are not interested in multi-occupancy buildings, short tenancies and intensive management. The following examples are exceptions.

Warwick Science Park, opened in 1984, was the first UK science park to involve speculative investment by a private funding institution. The initial site was made available by the university (14.5 acres) and the city council (9.5 acres) who hold the freehold. University of Warwick Science Park Ltd., UWSPL (see below, p. 38), took a 150-year head lease for the whole area. The first tranche of infrastructure was funded primarily by the former West Midlands County Council, with some assistance from Warwickshire County Council, the second and subsequent tranches of infrastructure by UWSPL, using receipts from premium payments for ground leases. Barclays Bank joined the project in 1982 and funded the development of the first building on the park, the Barclays Venture Centre, an incubation centre of 40,000 sq. ft. (3,716 m^2) carried out in two stages. This was a lease and leaseback arrangement. The bank is not concerned with lettings or management which are the responsibility of UWSPL. Subsequent phases of premises for rack rent have been backed by the West Midlands County Council, Coventry City Council and English Estates. The local authority funded schemes are also on lease and leaseback arrangements. English Estates are building 30,000 sq. ft. (2,787 m^2) of unit-based accommodation under licence, the intention being to sell on the head lease to an institutional investor once the building is fully let. In addition, seven companies have taken ground leases and constructed their own building: floor space in these varies from 4,000–40,000 sq. ft. (372–3,716 m^2). The initial development plan for the park anticipated that the twenty-four acres would be completed in ten years: it was in fact completed in four years. In 1987 UWSPL acquired from the city council a further eighteen acres (also on a 150-year head lease) which entailed buying out and relocating a rugby club on a long lease.

At South Bank Technopark in London, Prudential Assurance provided the finance to develop land made available by the local authority for Technopark 1. Phase 1 was opened in December 1985 and Phase 2 in September 1987. The Prudential own the land and buildings and have a fifteen-year management agreement with South Bank Technopark Ltd. (SBTL) which is responsible for lettings and management. There were external social factors which acted as a catalyst for the participation of the private sector in South Bank: the inner city riots of 1981. The Prudential could see the possibility of its property investment disappearing overnight.

At the Cambridge Science Park, Investors in Industry (now 3i) took a ground lease and developed incubator units on a two-acre site in 1983. It was later sold to the Eli Lilly Group.

At Silwood Science Park, 3i Commercial Property Ltd. has acted as developer and funded all development costs. The freehold has been owned by Imperial College, London, since 1947. Accommodation is made up of a Technology Transfer Centre (TTC) of 23,900 sq.ft. (2,220 m^2) and six other buildings totalling 45,900 sq.ft. (4,264 m^2): these were constructed under a building agreement and opened in 1987. The TTC remains in the ownership of the college; the other buildings are held on a ground lease with an option to claim the freehold. These buildings and their tenants constitute a typical 'institutional' package which 3i is likely to sell on to a financial institution – thus making Silwood Park unique. Such a development strategy is likely to be appropriate only in areas of high demand.

Since most second generation science parks have been in areas of high unemployment and de-industrialisation, it is not surprising that the private sector has shown little interest. The amount of unused or under-used industrial floor space in the Midlands and north increased during the early 1980s. Rentals have been low and showed virtually no growth while costs have risen. Thus, commercial investors in industrial property have tended to ignore development north of Watford. High technology industrial companies have preferred to locate in the south-east where the pool of labour available and the quality of life are expected to be better than further north. In most cases such companies have not seen fit to invest in their own buildings on science parks. By February 1988, only twenty-eight tenant companies had invested development finance in their own building.[1] At that time, there were thirteen ground leases on the Cambridge Science Park and seven on the Warwick Science Park. The proportion of the total investment (Fig. 3) is high because these premises are very much larger than the typical science park premises. Rapidly growing companies prefer to keep their assets in liquid form and to keep options open so far as future premises are concerned. Such an outlook has not been consistent with customised buildings in unpopular locations.

Wavertree Technology Park, Liverpool, is an interesting early exception. Although claiming to be a technology rather than a science park (see above, p. 12) and having no higher education institution involved as a partner, it has an on-site management team and a concern with encouraging new businesses that could arguably make it a science park. Plessey plc funded the first building to be occupied on the park: in 1984

Plessey Crypto took over 40,000 sq. ft. (3,716 m^2). The following year Powell & Scholefield moved into 25,000 sq. ft. (2,323 m^2), also on a 125-year ground lease. Both firms had moved from elsewhere in Liverpool. Wavertree is a joint initiative of the former Merseyside County Council, Liverpool City Council and Plessey and involves English Estates as developer of small- and medium-size units (totalling 62,000 sq. ft./5,760 m^2) and of the amenity building (medical and dental centres, bar, restaurant and bank).

The stimulus for Wavertree was the 1981 Toxteth riots which brought the then Secretary of State for the Environment to Liverpool. The need for a focus for the new industries of the future was recognised and a site was identified: sixty-four acres of former railway sidings, then derelict and land-locked. Plessey, Telecommunications and Office Systems Ltd. already occupied ten acres adjacent to the park. Wavertree was thus exceptional in two respects: it had an industrial company investing development finance in the north at the depth of a recession and it had the same company providing business support and technology transfer for the scheme as a whole (see below, p. 72).

Belasis Hall Technology Park, Billingham, is an initiative of ICI which has put up a third of the infrastructure and landscaping costs for the first fifty acres of this 167-acre site, the rest being provided in equal parts by English Estates and government (mostly central government) grants. It is the first park to be initiated by an industrial company and to have an industrial company fund the infrastructure in general rather than just the development of its own building. Development of Phases 1 and 2 has been funded by English Estates. Phase 3 will be carried out by Foster-Church, who will take a 125-year ground lease to develop a business centre totalling 100,000 sq.ft. (9,290 m^2). Units will range from 200-2,500 sq.ft. (18-232 m^2). The centre will include secretarial services, conference facilities and a restaurant. (See further below, pp. 52, 57 and 71–2.)

Emersons Green Science Research Park, near Bristol, is an ambitious proposal promoted by the private sector, albeit with the active encouragement of all the relevant local authorities. The south-west has, to date, no science park, being an area where local authorities have not seen the need for such investment since the economy is relatively buoyant, and where English Estates have little commitment. This proposal, announced in mid-1987, involves both a property company, Chesterfield Properties plc, and a subsidiary of Hambros Bank, Hambro Linsandro Ltd., together with Bath University, Bristol University and Bristol Polytechnic and a locally-born entrepreneur, Mr Christopher Turner, whose idea it was. Development funding will be raised from a variety of sources. The Emersons Green Development Company (EGDC), set up by Chesterfield, Hambros and Mr Turner, now includes the three academic institutions which have taken a fifteen per cent aggregate shareholding.

The scheme being formulated by EGDC is expected to cost around £250m and to cover some four hundred acres in all. There will be over 2m sq. ft. (185,800 m^2) of research and development and manufacturing space, including an academic centre of fifteen acres which will give the three academic institutions an on-site presence. Apart from the actual research laboratories occupied by the academic institutions, the academic centre will also have start-up units, campus housing for researchers and visitors, offices for professional and other firms to provide one-stop social facilities (particularly for the young new companies on site), meeting rooms and other accommodation. The centre will be run by the Science Research Foundation, a charitable body set up as a company limited by guarantee. The Foundation will be responsible for attracting funds to the centre. The scheme will also have about fifty acres of high-quality housing, mostly for workers on the site, a 200-300 bed hotel, with conference facilities, and some local shopping.

Strategies for Development

A variety of strategies has been adopted in order both to assemble resources (physical, financial and professional) for development and to enable suitable management expertise to be available and paid for. Although each science park may be considered unique the strategies fall into three groups (Fig. 4, see overleaf).

University Led and Funded

In these cases, the university has established the science park, using its own or borrowed finance, and has continued to be responsible for all development and management functions. Brunel, for example, raised funds by selling a ground lease to the International Tin Research Council. The finance thus raised, together with a loan from the Midland Bank, enabled further development of premises for rent. All the sites in this group are in areas where land prices and rents are high. Such a strategy is only possible where a university holds the land and either itself has substantial financial resources or where the demand for premises is high and outstrips supply. It will be noted (Fig. 4) that all the parks in the first group are in areas more favoured economically. In these areas there is both less incentive for local authorities to invest in job-creating ventures and more likelihood of industrial companies taking ground leases.

The Cambridge Science Park has shown a very different approach to funding from its successors. Early development, 1971-8, was funded entirely by Trinity College from internal sources. Development was slow and cautious and until 1978 all units were built for individual tenants. Even the blocks of small units (1,300 sq. ft./121 m^2) built between 1978

Fig. 4 Development of UK science parks (to end 1987)

Strategy	*Science park*	*Date of opening*	*Area (acres)*	*No. of tenants*
University led and funded	Cambridge/Trinity	1972	130	60
	Heriot-Watt, Edinburgh	1972	56	31
	East Anglia, Norwich	1984	12	4
	Surrey, Guildford	1984	70	33
	Sussex, Brighton	1985	3.5	12
	Brunel, Uxbridge	1986	6.4	20
	Cambridge/St. John's	1987	19	27
Joint venture companies	Merseyside, Liverpool	1982	2	17
	Aston	1983	22	45
	Manchester	1984	15.5	15
	Chilworth, Southampton	1984	26	20
	Warwick	1984	42	40
	Newtech, Clwyd, Deeside	1985	5	13
	South Bank, London	1985	1.7	53
	Birmingham	1986	8	9
	Cardiff	1987	1.5	4
	Keele	1987	15	8
Co-operative ventures	Listerhills, Bradford	1983	11.75	46
	West of Scotland, Glasgow	1983	61.5	21
	Springfield House, Leeds	1983	1.96	8
	Newlands Centre, Hull	1984	3	17
	Loughborough	1984	3	16
	Highfields, Nottingham	1984	18	23
	St. Andrews	1984	.74	4
	Aberystwyth	1985	6	6
	Mountjoy, Durham	1985	2.7	16
	Antrim	1986	74	4
	Mentec, Bangor	1986	2	3
	Bolton	1986	2.5	16
	Kent, Canterbury	1986	10	3
	Stirling	1986	14	10
	Swansea	1986	3.4	14
	Silwood, Ascot	1987	7.5	6

and 1982 were mostly pre-let. There was a number of reasons for this, relating partly to the status of Trinity College as an eminent and historic academic institution; partly to the financial standing of a college with a very large property portfolio; partly to the climate of local opposition to the expansion of Cambridge; and partly to the fact that this was an entirely untried form of property development in Britain. By proceeding slowly the college was able to ensure that it could control the nature of development and the type of tenant. It was also able to allay the fears of those concerned about the threat of 'dark satanic mills'.[2] Trinity was able to proceed slowly since it has owned the land for centuries and was therefore not incurring interest payment. There were virtually no opportunity costs either, since the land for what became Phase 1 was mostly derelict having been used as a tank base during the Second World War. Furthermore, planning applications for other uses (residential and industry) had met with no success.

Subsequently, the college has funded development by selling ground leases (99 or 125 years). The first of these was that of Cambridge Consultants in 1977. This building was the last for a tenant company in Phase 1. In the remaining phases, buildings have been financed either by the college or, in the case of ground leases, by the lessee. Access roads, communal landscaping and the social centre (completed in 1984) have been funded either by the college or by in-kind payment or by ground lessees. This incremental approach has meant the college has not itself had to go outside for development funds. Nevertheless, establishment of the Cambridge Science Park was very slow: it took a decade to create its present image (see above, pp. 7–8), despite benefiting from its address.

The Surrey Research Park, Guildford, has been developed in a more pro-active manner. The land is owned by the university through a charitable trust. It was purchased in the 1960s with public subscriptions. Thus, like Cambridge, it is held at a relatively low historic cost with no interest charges. Unlike Brunel Science Park, where the Treasury 'clawed back' fifty per cent from the sale of the first ground lease and may in future take more, the Surrey Research Park was not bought with any government money and the Treasury therefore has no claim on income from it. The university has funded the infrastructure and landscaping of the whole site, the development of Chancellor Court, five buildings comprising 47,000 sq. ft. (4,366 m^2) in Phase 1 and six buildings comprising 62,200 sq. ft. (5,760 m^2) in Phase 2, by borrowing and the sale of ground leases.

The first building to be constructed was the Grand Metropolitan Innovation and Development Centre (GMIDC) financed by Grand Metropolitan plc. GM took a 99-year ground lease on annual payment terms. The GMIDC has now been bought by the university and renamed the Surrey Technology Centre. BOC and BOC Cryptoplants took 125-

year ground leases on an initial premium payment and peppercorn rental terms. The BOC buildings have thus provided the research park with substantial capital. The park has also built up a good credit facility with its bankers. The university will develop the remaining area inside the loop road probably on design-and-build contracts and will let premises on rack rental terms. The area outside the ring road is expected to be developed in units of 30,000-100,000 sq. ft. (2,785-9,290 m^2) on design-and-build contracts for specific occupiers on long leases of up to 125 years. This way the university will reap the development profits.

Joint Venture Companies

Here separate legal entities have been set up. The financial and other commitments between the parties vary from one science park to another. The company can act as the client for which the development is carried out by one or more of its partners. After the initial phase of development, the company itself may act as the developer. The joint venture company provides a vehicle through which control and management as well as future development may be secured and co-ordinated. This mechanism ensures a continuing positive role for the university and other parties in the company.

Aston Science Park is run by Birmingham Technology Ltd. (BTL), a commercially orientated private company limited by guarantee with no share capital. The original partners were Aston University and Birmingham City Council. Lloyds Bank is also involved (see below, p. 52). Development funding has been entirely through the city council, either directly or using Inner Area Partnership Funds (Phases 1, 2 and 3) or European Regional Development Fund assistance (Phases 2 and 3). BTL was the client for Phases 1–3, when Birmingham City Council provided the development team. In future phases, BTL will be the developer but will continue to involve the city to secure the benefit of grant aid for infrastructure and reclamation work. The company is also very actively involved in managing the park (see below, p. 61).

Warwick Science Park is run by the University of Warwick Science Park Ltd. (UWSPL), a company with four partners. The university holds a 35 per cent share. The other partners are Coventry City Council with a 45 per cent shareholding, the West Midlands Enterprise Board,[3] formerly West Midlands County Council, with 11.5 per cent and Warwickshire County Council with 8.5 per cent (see pp. 32 and 61). UWSPL carries out both development and management functions.

Manchester Science Park is run by Manchester Science Park Ltd. (MSPL), formed in 1983. The partners are Manchester City Council with a 35 per cent holding, the University of Manchester with 22 per cent, UMIST with 13 per cent and four companies (Ciba Geigy plc, Ferranti plc, Fothergill and Harvey plc and Granada Television Ltd.)

each with 7.5 per cent. The city council has funded and carried out the development of Enterprise House (Phase 1) and provides grant support to industrial companies. The development of Phase 2, Pencroft House, is being co-ordinated by a development company using an urban development grant. The university provides access to university expertise and facilities. MSPL has taken a 125-year head lease on the fifteen-acre site from the city council. The company is managed by a chief executive (see below, p. 62).

Co-operative Ventures

This is the most popular strategy. It does not involve any separate legal company and is thus highly flexible. Typically, a local authority or development agency takes the lead role, depending on who provides the major financial input. Compared with those already outlined, this is the strategy with the least input from the academic institution, which will normally work through its industrial liaison officer. In some instances an on-site manager is provided by one of the partners.

Listerhills Science Park at Bradford University is a joint venture between the university, the city council and English Estates (see above, p. 31). The university provides support and the services of its director of industrial liaison. The city is generally supportive, gives grants towards research and development costs and helps formulate development proposals. Management is informal (see below, pp. 58–9).

Bolton Technology Exchange is a joint venture between Bolton College of Higher Education, Bolton Metropolitan Borough Council and English Estates. The initiative came from the college but was given immediate support by the borough. A feasibility study was commissioned to consider the viability of a science park in Bolton. The recommendation was positive and developers were approached, including English Estates. Towards the end of the construction period (i.e. late 1985) a commercial manager was sought (see below, p. 60).

Evaluation Aspects

Market Appraisal

The initial feasibility study must identify the potential market segments. It should include an analysis of the local economy, its composition, linkages and growth potential. The strengths of the local economy will be the most likely basis for the early growth of the science park and it is important to identify what will attract potential tenants. The profile of tenants will vary for each development.

Small existing firms form the majority of science park tenants and for them the length of the journey to work is significant. Managers of young and small firms do not move away from their industrial contacts; nor do they wish to uproot their family or that of their key workers. Warwick Science Park is particularly fortunate in being easily accessible to a large population. Durham, on the other hand, lying as it does between Teesside and Tyneside, has a very small 'journey to work area'. In most of the 1980s science parks, it has been local firms, looking 'over the fence' and seeing the possibility of a better address and image on the science park, who have formed the first tenants (see below, p. 68).

Research units of large organisations and fast-growing medium-size firms from other parts of the country are likely only to be attracted to well-established sites, especially outside the south of England. Durham is an exception in this respect.

The results of the market research will need to be compared with the initial objectives of the proposed development. It may be that the time is not right or that more effort will need to be put into creating a demand where none exists. Listerhills in Bradford showed that a successful science park could be created where no demand was identified (see below, p. 51). Existing science park managers stress the importance of market research and the need to design the scheme for the potential tenants. A strategy for promotion will be essential (see below, p. 53).

Site Identification

The location needs of science parks (see above, pp. 20 ff.) have meant that land availability is a major constraint, especially in the light of the low investment returns anticipated in many locations. Almost all sites have been wholly or partly in the freehold ownership of the higher education institution or local authority. Thus the land could be made available at low cost, a significant factor in development appraisal. Some of the inner city sites may, however, inhibit future growth owing to their size; perhaps, in the early days, horizons were too narrow. The next stage of growth is already a matter of investigation for many sites in the north, e.g. Listerhills.

Planning Permission

The granting of planning permission has been a problem for sites in areas with restraint policies. There have been restrictions on manufacturing processes and on office use. Also planning authorities have retained a tight control by granting permission in phases in accordance with detailed planning and design briefs. South Cambridgeshire District Council and Guildford Borough Council were both concerned about the effects of additional employment. At Cambridge (now 130 acres) a section 52 agreement[4] has been used to restrict development to science-based industry. There have been five separate outline permissions, the sixth application (for Phase 5 which will complete the site) being submitted early in 1988. Each building has been the subject of a separate detailed application. At Surrey (sixty-eight acres), five phases were identified in the initial development brief, outline permission being granted for the first phase of fourteen acres in December 1983. The second phase for which permission was granted, also fourteen acres, was almost completed at the time of writing. The addition of Phase 3, 7.2 acres, for which an application was submitted in March 1988, will complete the area within the ring road. At Silwood Park, Ascot, the outline planning permission restricts use to 'purposes complementary to the activities of the Imperial College of Science and Technology and shall be deemed to include the carrying out of any process or activity of which the research, design and development are a fundamental and primary part, including scientific control and experimentation'. This is in order to enable the Royal Borough of Windsor & Maidenhead 'to maintain control over the use of the buildings and to comply with Green Belt Policy'.[5]

In most areas planning permission has not been problematic. It has often been possible to get a wide-ranging permission. At Listerhills, for example, permission was granted for (old Use) Classes II (office), III (light industrial), IV (general industrial), and X (wholesale warehouse).[6] In several cases, developers have obtained permission for science-based industry or research and development but have negotiated a fall-back arrangement such that should the accommodation not be fully let within two years of completion some other use would be considered favourably. This was the case in Durham, for example, where offices would have been a lettable alternative.

In future, most science parks expect to obtain permission for (new Use) Class B1 (business).[7] Those local authorities concerned about possible impact on city centre offices, industrial areas or employment levels may well try to restrict uses more narrowly to research and development and associated activities by granting *sui generis* permissions. In the longer term, however, local authorities will need to rethink their policies and review local plans in the light of the Use Classes Order 1987. Given the specific objectives of science parks and the active management thereof, the impact of the change to Class B1 will be small compared with that in business parks and high technology industry generally. The effects of the Order on rental and site values will depend on supply and demand in each area. In general, the increased flexibility should be to the advantage of science park developers.

Development Appraisal

The results of the market appraisal need to be interpreted before a decision is made to proceed. The development proposal must be matched with an identified demand or with a deliberate strategy to create demand. Imagination, creativity and flexibility will need to be applied to all aspects of the proposal: to timing, funding, premises and management. The scheme must be seen as a long-term investment in terms of financial return as well a technology transfer and job creation. It is interesting to note, however, that in virtually all cases demand has been far greater than had been anticipated. Where voids of up to two years had been allowed for, premises have frequently been fully let within a year (see below, p. 53).

Property-related decisions, speculative units or pre-lets, length and terms of leases, sources and methods of financing, are all important. Equally important at the development appraisal stage are considerations of mechanisms for mobilising the university and other partners' non-property resources, for strengthening the commercial aspects of young companies, for access to venture capital and for managing the scheme. Effective marketing of the development (see below, p. 53) is crucial to reduce voids.

Funding

Development funding (Figs. 1 and 3, see above, pp. 29 and 31) for both infrastructure and buildings has been a major problem for UK science parks. In addition to the uncertainty of a new market, there are two further reasons for this: the rate of development and management costs. Development is usually phased over a relatively long period owing to the entry criteria for tenants and the small amount of development per annum, especially at the beginning. Thus, land and infrastructure costs are unlikely to be viable on conventional commercial criteria. At the same time, intensive management is an essential prerequisite of a successful science park. This management which is seldom self-supporting in the early years is expensive – much more expensive than conventional estate management (see below, pp. 57ff.). Thus there is a need for participation by financial partners able to adopt long-term horizons and for investment finance for management as well as development costs.

In future, other sources of development funding will need to be tapped, both for new initiatives and for the continued growth of existing schemes. Since science parks are now an accepted form of development and since they have grown rapidly showing good rental and capital growth, it is appropriate to examine the means by which private investment could be attracted.

Since 1983, the investment market for industrial property, i.e. insurance companies and pension funds, most portfolios of which have been built up only since the early 1970s, has reduced its involvement in industrial schemes. During 1987, however, performance improved significantly and since the stock market crash (October 1987) the institutions have begun to increase their industrial property portfolio. These institutions are looking to earn an income and obtain capital growth. Not only have returns on property assets been low compared with other markets, but returns on industrial property have been outstripped by offices and even more so by retail in every year since 1980.[8] In order to involve institutional investment, it will clearly be necessary for science parks to present themselves in a manner which is attractive to those institutions, whose criteria tend to be inflexible.

The institutions are seeking clearly identifiable properties, in single ownership with leases which require the occupier to undertake the management and maintenance. Such unencumbered properties are relatively easily marketable to similar types of investors, most of which have few management resources. This kind of property does not match that of most science parks, though Silwood Park is a notable exception (see above, p. 33). The problem is aggravated by the absence of a good secondary market. In order to make science park investment attractive, it will be necessary for an overriding lease to be taken by an intermediary between the science park tenant and the institution. This lessor will need to have proven competence to manage the property in such a way as to ensure a relatively risk-free asset.

The institutions are also seeking evidence of future profitability. Although the circumstantial evidence from science park developments looks good, the relative newness of the market means historical evidence is thin. This factor, together with the uncertain covenants offered by most tenants, is clearly unattractive to investors. In this context, the study initiated by UKSPA, in February 1988, to present a record of the investment performance of science park properties will be welcome. It is being carried out by an experienced team of private and public sector individuals.[9] In addition to the factual evidence they will identify the most suitable development and investment routes, including financial vehicles, to encourage greater private sector participation; the timing and extent of any public sector involvement required; and the most suitable forms for collaborative public/private development and investment.[10]

Property development companies are showing more interest in science park development than the financial institutions. They also, however, tend to prefer an overriding lease to be taken by a body, possibly a local authority, which would bear the risks and costs of flexible and short leases. Development companies are also inclined to look for support from the public sector with infrastructure and land costs. A conflict of interests in likely here: the development company may want control while many higher education institutions will be reluctant to relinquish their overall role.

The commitment as well as the resources of existing agencies will need to be re-examined. Some local authorities, such as Birmingham City Council in Aston Science Park, have so far shown strong commitment. This development has been a major initiative for Birmingham. Nevertheless, there will come a time (perhaps not far away) when Aston Science Park will have to seek development funds elsewhere. Other local authorities such as Manchester and Nottingham have regarded their science parks as only one of numerous initiatives in those cities.

Higher education institutions are also subject to a range of different pressures which result in conflicting objectives. Often, the science park is just one of many ventures designed to give increased prominence and relevance to the institution in the local community and within industry. Frequently, the institution will be interested in 'seed-bedding' from its own intelligence and attracting research organisations, but it may not wish to be overwhelmed by a large-scale successful science park. For many institutions the effects on economic and environmental regeneration of the inner city will be an added coincidental bonus rather than a true objective.

The way forward is undoubtedly through joint private and public sector ventures. In many instances the risks of science park development are now sufficiently reduced to enable private developers to become involved without sacrificing their normal commercial criteria. Development companies and financial institutions could well participate but their objectives and constraints as well as those of the public sector partners will need to be articulated and conflicts resolved. It will be important also that effective management structures are created. New funding vehicles such as unitisation will not be an immediate solution, the yields will need to be right first.

References

1. Rowe, D., 'Achievement from enterprise', UKSPA Conference, February 1988.
2. A term used in the Holford Plan in 1950 which formed the basis of the Cambridgeshire Development Plan approved 1954, the First Review of the Cambridge Town Map approved 1965 and the Preliminary Report for the Development Plan Review in 1967.
3. The West Midlands Enterprise Board Ltd. (WMEB) was established in 1982 by the West Midlands County Council (WMCC) as part of its economic strategy to supply a gap in the provision of long-term risk capital. Initially, it focused on medium-sized manufacturing firms. Since the demise of the metropolitan counties (1 April 1986) the WMEB has continued as a locally accountable company. It has widened its activities and has taken over some of the responsibilities and assets of the former WMCC, including the shareholding in UWSPL.
4. Carter, N., and Watts, C., *The Cambridge Science Park*, Planning and Development Case Study 4, Surveyors Publications, 1984, pp. 18–19.

5. Royal Borough of Windsor & Maidenhead, Notice of Planning Permission, 28 August 1985.
6. Town and Country Planning (Use Classes) Order 1972, S.I. 1972 No. 1385.
7. Town and Country Planning (Use Classes) Order 1987, S.I. 1987 No. 764.
8. Investment Property Databank.
9. Debenham Tewson & Chinnocks, Legal & General, English Estates, Department of the Environment, Investment Property Databank and UKSPA.
10. Rowe, D., op. cit., February 1988.

5. MANAGEMENT OF SCIENCE PARKS

Science parks require positive, confident and reliable management. With any property there are liable to be aspects which are unpredictable and which require more flexibility than is needed for many businesses. Compared with typical industrial and commercial users, science park tenants change their property requirements more often and more rapidly. They seem to need enlarged premises almost overnight. This is owing both to the cyclical nature of the activities within small companies (see below, pp. 67ff.) and to the rapid expansion of high technology companies (see above, pp. 10–11). It is therefore necessary for management to be close enough to the tenants to be able to recognise their needs almost before they do. This means frequent and regular personal contact and understanding the tenants' business activities and expectations. It calls for knowledge and proximity on the part of management: being on site, accessible, approachable, interested and aware.

Objectives

It is apparent from the above that science parks will require management skills over and above those of traditional estate management. It is also apparent that, owing to the importance of non-property objectives and to the extent of public sector involvement in many science parks, the normal commercial criteria for measuring property performance and investment yields will not be adequate. The public sector will have social and economic (as well as financial) objectives which mean the inclusion of non-financial criteria in investment decision-making.

It is important to clarify the 'needs' with which management will be concerned: property, finance, technological/scientific or business advice. Clearly there is an interrelationship between property needs and the rest. A company in financial trouble will be a bad risk and at some stage may be unable to pay the rent. Such financial hardship may be the result of inadequate business planning but could, for example, be due to technical difficulties which require access to particular testing equipment, inadequate funds to develop new ideas through to a marketable product or lack of finance to purchase necessary equipment. It is, therefore, in the interests of good property management to ensure that the other needs are taken care of.

Property management involves more than collecting rents and negotiating rent reviews. It may be defined as follows:

> Property management seeks to control property interests having regard to the short and long term objectives of the estate owner and particularly to the purpose for which the interest is held: to negotiate lettings and to indicate and negotiate rent reviews and lease renewals, to oversee physical maintenance and enforcement of lease covenants, to be mindful of the necessity of upgrading and merging interests where possible, to recognise opportunities for the development of potential and to fulfil the owner's legal and social duties to the community.[1]

Such property management will demand control over tenants where, for example, their behaviour contravenes the lease or would be a nuisance to neighbours. Science park management, on the other hand, will also entail the promotion and marketing of tenant companies in order to fulfil the non-property objectives of encouraging technological transfer and creating employment and wealth in the local community. Science park management will thus find itself acting not only on behalf of the landlord, but also on behalf of the tenants. Any property manager occupies a crucial role in the relationship between landlord and tenant but in science parks it is much more of a two-way activity.

Management Aspects

Technology Transfer

An important objective of science parks is to reduce the time normally taken to transfer the fruits of scientific knowledge into commercially viable products, i.e. speeding up the process of technology transfer. This process is complex and the factors which determine its speed and success highly variable (see below, p. 73). Nevertheless, the management of most science parks includes responsibility for facilitating technology transfer. The priority it is given and the way in which it is tackled vary considerably.

It may be argued that the very presence of incubator units in a science park close to the university will encourage academics to venture into the commercial world. It is improbable, however, that the mere presence of accommodation in itself will initiate the process. Those entrepreneurs at the right stage in product development may indeed take up premises. If accommodation is in short supply the presence of space near the university will encourage new companies to stay near by. The fact that youthful companies are close to the parent university and able to take advantage of supportive services in 'sheltered' science park accommodation, may indeed ease the maturing process of a firm, but is likely to have relatively little effect on whether it is established in the first place.

'Short circuiting' the process of technology transfer requires bringing together those people making products on a commercial basis and those working in academic research in related areas of knowledge. This has traditionally been done by industrial firms who initiate contract research. It normally involves a relatively long-term relationship through, for example, sponsoring a particular project or a series of PhD students. The results may or may not be commercially viable but they tend to be slow in realisation. Another approach which is likely to yield more lines of contact and thus a higher or faster success rate is by industrial firms putting research units on to a university campus so that they are in close contact with the relevant university department e.g. the Leicester University/ICI Joint Laboratory in the Biochemisty Department. Science parks are able to assist by providing space adjacent to the university campus for such units. Alternatively, the science park may accommodate the university's own research units.

The Mountjoy Research Centre at Durham University is interesting in this respect. The first tenant was UDIRL (the University of Durham Industrial Research Laboratories) occupying 6,280 sq. ft. (583 m^2). This is an association of some seventy academic staff in several research areas but mostly associated with contracts having a chemisty bias or requiring instrumental analysis and interpretation. It is self-financing and has its own links with industry. Equipment has been provided partly by industrial suppliers and partly through university funding. UDIRL is seen by some as a more appropriate place for contract research than individual university departments. It is separated from the more fundamental research which needs to be free from commercial constraints; at the same time researchers who might normally be isolated in separate departments are brought together. Some departments, however, believe that contract research is more appropriate within individual departments where it runs alongside the longer term fundamental research.

These opposing views as to where contract research should best be located each have merits which need to be understood by those involved in science parks. It was the belief that contract research is incompatible with basic research and teaching that led, for example, to the setting up of the Leicester Biocentre, a contract research unit which grew out of the Department of Genetics at Leicester University and which is located on the campus. These inherent conflicts will doubtless become more apparent as central government forces through its policy of making universities and polytechnics more commercially orientated.

The Cambridge Science Park has recently let eight acres to GEC, who will occupy a building of some 100,000 sq. ft. (9,290 m^2) together with an adjacent 'clean room' building. London International moved its research and development facilities, formerly adjacent to its main manufacturing site in Essex, into a 16,500 sq. ft. (1,533 m^2) building there in 1987. The new site and building were chosen in order to develop linkages with academics in the same field of interest and to create a scientific environment which will attract high calibre staff and provide a solid basis for long-term technical growth. In 1981, the University of Cambridge Engineering Department set up a Microelectronics Research Laboratory on the park in order to extend the interaction between industry and microelectronics research at the university.

Technology transfer should also be encouraged by a closer network or community between science park tenants, industrial firms elsewhere and university academics. Surrey University has recently established SUNET (the Surrey Network for Industrial Collaboration) to encourage liaison between industry and the university and to provide a means by which specialist research equipment and scientific and technical consultancy can be supplied. SUNET has a dozen members who are large companies paying substantial subscriptions. All tenants of the Surrey Research Park are automatically given free membership. This entitles members to free use of the university library, sports facilities and car park, and to attend scientific and other technical meetings normally restricted to staff and students, as well as seminars set up especially for SUNET members.

Tenant Support

A requirement for tenant support is implied in the objectives of science parks but is also in the best interests of property management. A successful park is likely to expand and to increase income and capital values. Future industrial growth and thus future demands for industrial property will most probably come from the new technologies and the younger firms. Thus it is in the property manager's interests to understand the needs of small firms and to nurture them in order to achieve rental growth. Put another way, this means growing your own firms to create demand. It means managing a tenant portfolio to maximise industrial potential as well as managing a property portfolio to maximise the owner's objectives. Tenant support may be seen as the non-property aspect of management: it is essential that this is approached in an active and positive manner. Such support is an important factor in the low failure rate of science park firms. During 1987, it was nearly four times lower than the national average for the small firms sector.[2]

Many tenants, especially the small and newly formed companies, do require advice and support. Just as such tenants appreciate a single charge which covers all accommodation costs, they also appreciate being able to get advice on various business matters (at least in the first instance) from one source. Small high technology firms typically concentrate on the technological aspects of their business neglecting both marketing and financial management. In the early stages of the growth of a firm (see above, pp. 10–11), it often lacks expertise in these matters. This is borne out by a survey[3] in 1986 which showed that 51 per cent of science park firms found time constraints on the management team a major problem in their first year of operation. Thirty-three per cent found obtaining finance and 37 per cent obtaining payment from large debtors a problem. Furthermore, even though a firm may survive without management support, such support will undoubtedly foster faster growth. The popular supposition that the British are good at invention but not at the commercialisation of ideas is supported by evidence which shows that a major cause of the failure of new products is inadequate market awareness and marketing competence rather than technical problems.[4]

The support required by individual tenants at a given time will clearly vary. Science park managers must know the tenants well enough to sense their needs and problems before they do, whether the problem is one of requiring enlarged premises immediately, not being able to pay the rent or not having a market for a new product. Such knowledge will be acquired in a variety of ways. It will mean being on friendly terms but also being known to give a professional service, advice that can be relied upon. It means being sociable but respecting privacy and being able to distinguish between the different reasons for a firm needing privacy; it could be because they are very busy or because they have no orders. Many science park managers, including those operating almost single-handed, appreciate well the need for 'management by walking about'. Those who approach management of science parks from an ivory tower are not optimising their assets. Science park management needs to be seen to be running a successful business, providing a facility which tenant firms see as worth paying for.

In areas where there is high demand for accommodation and rents are high it is likely that few start-up firms will be able to afford the rents. In such a situation the cautious science park manager is more easily able to select tenants with conventional good covenants. These sites, however, are few and it can be argued that such an approach to tenant selection is not consistent with science park objectives. In most science parks tenants will from time to time need help in preparing their business plan or marketing strategy. They may also have difficulty in raising finance.

Local enterprise agencies are playing an increasingly important role in encouraging business initiative. From their common core of business counselling, some have focused on technology and innovation, some on information and advice, others provide investment finance. Subscribers include many of Britain's large companies which give resources in cash and kind, for example the secondment of qualified staff. The umbrella organisation Business in the Community co-ordinates links between the private sector and job creation, and an annual government grants system now exists. The European Business and Innovation Centre Network (EBN) acts as a clearing house for the exchange of information and experience between various business and innovation centres being set up throughout the European Community. As well as the centres themselves, members include banks, private companies, nationalised industries and consultants. The aim of these centres is to help new small and medium-size enterprises develop innovative products and processes, generally using new technology. They are intended to be complementary to other enterprise agencies. Many science park managers see benefits from participating in such local initiatives. Some such as Aston are members of EBN.

Business advice Business advice may or may not be given by science park management itself. It may be seen as requiring an intensive input in the first year or two when incubator units (see above, p. 23) are being let for the first time. Subsequently, the tenants' needs will be less concentrated. After the initial period, a number of science parks have been able to satisfy tenant needs by encouraging a business management consultant to take up a tenancy on the science park. A similar approach is often used with regard to the provision of secretarial services. The business consultant or secretarial services find part of their work on the science park from other tenants and the science park management sheds an expensive responsibility. Frequently, the consultant or secretarial services will also do work for the science park management, perhaps in exchange for a concessionary rent.

Listerhills faced a particular problem in 1983: the feasibility study had found the potential market to be almost non-existent and the completed development remained unlet. A positive approach was adopted. Two staff from Bradford University were persuaded to move in to form a company (still there and very successful). A manager was seconded by English Estates (six months full-time then six months part-time) to devote his energies to marketing the scheme and to offer business advice (property and financial) to local companies. After about twelve months a critical mass was reached and almost overnight there was a waiting list for premises.

Finance As with any growing business science park firms need the right information and advice about sources of finance to suit their particular needs. At the start-up stage small amounts of seed capital are required to prove the viability of an idea. At the venture stage when prototypes are being developed, market research carried out and production processes being set up larger amounts are required. At the seed capital and venture stages there is little security thus the risks are correspondingly high. At the development stage when a proven product is being marketed and the business expanding, the risks are less. The number of venture capital companies has increased from twenty-seven in 1981 to 126 in 1986,[5] when £384m was invested in six hundred UK companies.[6] The total pool was estimated to be around £3b.[5] The emphasis of most funds, however, is on development rather than venture or seed capital. Management costs on small loans are considered prohibitive: £250,000, or more often £500,000, is a minimum. There is a particular need, therefore, for smaller amounts of risk capital, especially below £100,000 (see below, pp. 69–70).

In many cases, science park managers act as agents for economic development companies established by local authorities. The most well-known but by no means the only ones are those set up by the enterprise boards. For example: Worknorth Ltd., a subsidiary of the Greater Manchester Enterprise Board in conjunction with the Co-operative Bank, provides sums of between £10,000 and £100,000; the Warwickshire Venture Capital Fund and the Coventry Venture Capital Fund, which are subsidiaries of the West Midlands Enterprise Board set up in conjunction with the local authorities, provide sums of between £25,000 and £50,000. Such funds, however, are not specifically aimed at innovation or science park tenants.

Aston Science Park is unique in having its own venture capital fund. From the outset it was considered essential to provide venture capital under the control of the management company. This, it was anticipated, would place the science park in a unique position and help to overcome one of the major difficulties for new high technology based companies in the Birmingham area. The fund of £2m was contributed in equal parts by Birmingham City Council and Lloyds Bank and is administered by the science park company (BTL). Lloyds thus became the third partner in BTL. Income to BTL is generated from its investment in tenant companies. It has invested in twelve out of a total of forty-five tenants and has always included an element of equity participation (usually between twenty-five and fifty per cent).

Belasis Hall Technology Park Ltd. (BHTPL) also recognised at the outset that venture capital would be vital. A finance company recently established in London is planning to take premises on the site. It will concentrate on amounts up to a maximum of £500,000 and is backed by several financial institutions. The chief executive of BHTPL will assist in the preparation of tenants' business plans. It is anticipated that the fund will be in operation by January 1989. In addition, a joint initiative has been set up BHTPL, Barclays Bank and Price Waterhouse: Springboard. This is a financial management package available only to Belasis tenants and includes business planning and tax advice together with loan facilities.

Hambros Advanced Technology Trust (HATT) announced (February 1988) a special funding scheme, HATT Science Park Investments (HATTSPI), the first to be targeted specifically at science parks. The scheme has been devised with the assistance of UKSPA and will involve HATTSPI working with individual science park managers. This partnership will provide companies with an initial analysis and discussion about their business plans free of charge. Amounts of £20,000-£100,000 have been available from the £1m which HATT set aside for 1988.

Marketing

The standard approach of an on-site agency board and advertisements in the property press is quite inadequate. Market research will identify the potential market, its characteristics and its strengths (see above, p. 39). These should be built upon. The next step is to highlight the special features of the proposed development which will attract potential tenants. With this knowledge a real demand can be created from that potential market which is unlikely for the most part to have given any prior thought to locating on a science park. Although there is a recognisable science park property sector at a national level, each new development is breaking new ground locally.

The absence of a local high technology base has meant that some science parks (e.g. Bolton), which were set up to regenerate an area by bringing new technologies in to offset the declining industries, have needed to go outside the local area to find tenants – an approach which was adopted for Birchwood from the start. In such areas where the potential market looked thin (e.g. Bradford and Bolton), successful science park managers have seen tenant support as an important complement to active marketing. The business support service run at Listerhills by English Estates has become a model which other science park managers have seen the need to offer (see above, p. 51).

Industrial location decisions are not thought through rationally by most firms. Those which have given the matter any consideration typically look only at a short list of three or four sites. Even amongst those firms which do assess a range of sites, the final choice is often made for non-business reasons. This fact needs to be borne in mind in drawing up a marketing strategy. It is important for a new science park to be on that short list.

On the assumption that demand is unlikely to exist for an unknown product a high profile has been seen as essential by the more active science park managers, so that the next time a firm relocates it will give their site some consideration. Maintaining such a profile in terms of general advertising or direct mailing can be expensive for science parks operating on a slim management budget. Many partners have refused to spend money in this way, an attitude which has been shown to be short-sighted. The successful early letting of the first phase of Belasis Hall, where 50,000 sq. ft. (4,645 m^2) of units varying from 160–5,000 sq. ft. (15–465 m^2) in five separate buildings, were fully let to thirty tenants within nine months of opening, proves its value. The development appraisal had assumed a period of two years before it was fully let.

Three aspects which ought to be clear in any marketing are frequently matters of confusion; first, the benefits of being on the particular site, second, the nature of the property itself and, third, the person who should be contacted. The lack of clarity no doubt reflects the tensions and inadequate co-ordination between the partners in the science park development. It results also from a desire to indicate the range of initiatives being taken and the facilities being offered by the academic institution, but it reflects an inability to see the advertising from the prospective tenant's viewpoint. Science parks which are advertised as one of many developments by English Estates or a city council are far less likely to be noticed than those advertised individually under their own name. Provided the promotion is coherent, however, there are real marketing benefits to be gained from supportive partners who help to give continuous exposure to the existence of the science park. This has happened, for example, at Warwick Science Park where the local authorities and the university have assisted in creating a good image.

A number of science parks have built up their own database of firms in the locality in appropriate technological areas. This can be used as a basis for targeted mailshots. Advertising tends to be done selectively in, for example, the *New Scientist* or *Financial Times* rather than in the property press which has been found to bring little result. Few science parks put the marketing in the hands of estate agents. Even those which do employ agents, for example Surrey, draw up the strategy themselves, merely using the agent to carry out the work.

Lettings Policy

The criteria adopted with regard to tenant selection are a crucial element in achieving the objectives of the science park. The policy must be flexible enough to respond to unexpected demands, yet firm enough to resist unsuitable tenants. Most science park managements agree that it is important to set the tone by the first tenants and that one rogue can do inestimable damage, far outweighing the loss of rent which might have occurred by waiting for the right tenant. Suitable tenants will be not merely those who are clean and well-behaved. Such conventional criteria of estate managers for identifying good covenants as past profits record and full order books will not be appropriate for very young firms. The challenge in science park management is to attract and pick out small firms with strong promise in those knowledge-based industries which have potential both for linkage with academic institutions and research organisations and for the local economy in terms of job and wealth creation. There is a clear interrelationship between marketing and lettings policy. The tenant selection stage is a good time to identify areas where a firm may need support.

The explicitness of policies and the specific criteria vary. Some science parks see actual technological linkage with the university as essential, others are more concerned with firms which will grow. Some aim to restrict firms to particular industrial sectors which are complementary to areas of acknowledged excellence in the university; others want to import different expertise. Some require firms to be involved in research and development; others are less concerned about this.

Admission of tenants is usually a matter for a clear decision – affirmative or negative. There are however some grey areas. The most common is where the activity is mainly an office function, concerned with administration, sales or marketing; others may be where the activity creates an unacceptable environmental impact in terms of traffic or noise. Sites where demand is high will inevitably have less need to be concerned about less appropriate tenants.

At Aston Science Park, policy is based on the overall aim which is to create wealth and employment for the surrounding area. Tenant firms must be knowledge-based and have good growth prospects which can be accelerated by being on the science park. Tenants may benefit from the funding possibilities, from technological links to be fostered with the university and other tenants on the park, from business expertise (e.g. financial advice available on a consultancy basis), and business services (reception facilities, telex, etc.). Tenants are required to be engaged in activities to which the skills of the university or other tenants can make a positive contribution. Assessment of potential tenants is in two stages: first, consideration by the management team of the overall proposal; second, a technical report by an appropriate member of the university staff, with a market survey to attempt to verify the sales forecasts within the applicant's business plan. This process can take between four and thirteen weeks. Advice on grants available is also given.

Surrey Research Park tenants must be engaged in research and development (broadly defined) and not in manufacturing. In selecting tenants, the type of activity in which the company is engaged (research, prototype development, manufacturing, sales) will be examined, together with the employment profile (qualifications). If the company is trading (and most have been) some security deposit, banker's guarantee or similar, is required. In the case of start-up companies, a business plan is examined. In fact, there has not yet been such a company other than from the university and that has been viewed differently because it is 'known'. Linkage with the university is not mandatory: there is a preference for growth potential. Tenant selection is a joint activity between the research park directors and the agents.

Warwick Science Park tenants are expected to demonstrate a relationship (existing or potential) with work at the university, though this need not be in technology. A user clause in the head lease and all other leases restricts tenants to firms involved in research and development and manufacture essential thereto. This user clause reflects the terms of the planning permission.

South Bank Technopark is looking for business growth and people with drive rather than qualifications, but selects tenants which need help and would thus benefit from being in the Technopark.

Leases

As previously mentioned, flexibility and simplicity are important criteria for the typical new technology firm. Flexibility means easy-in/easy-out terms; simplicity means a single accommodation charge, rather than separate charges for rent, rates, services, insurance, etc., and perhaps a tenancy agreement rather than a lease. The use of periodic tenancy agreements means simplicity for the science park management as well as the tenant. No solicitor or property manager's expertise is required (Manchester, Bolton and South Bank work thus).

Some science parks have been able to use typical institutional leases on full repairing and insuring (FRI) terms for twenty-five years, notably Cambridge, Surrey and Silwood. Others have taken a very different approach, in particular with regard to external maintenance, repair and insurance and length of lease. Shorter leases with landlord responsibility for all external finishes and areas give advantages to the landlord as well as the tenant. Control and enhancement of the development as a whole is achieved more effectively and the tenant portfolio can be reviewed more frequently and those in incubator accommodation persuaded to move on. Unsatisfactory tenants can nevertheless survive where the Landlord and Tenant Acts operate.

Typically, leases are available for six, nine or twelve years with three-yearly rent reviews. Three-year tenancy agreements breakable with three months' notice are frequently offered. In some cases, three- or one-monthly licences may be available for very small units or even rent-a-desk facilities.

The structure of rent and service charges varies considerably, both in terms of items included and the calculation and charging arrangements. The variation occurs between different science parks and between different buildings on the same science park.

Aston Science Park leases are taken outside the Landlord and Tenant Act 1954. No lease is longer than five years to enable a review of occupation and rent to take place. The service charge in the Phase 1 incubator units includes a comprehensive range of services and is designed to free tenants of property and reception worries so that they may concentrate on their business. The charge includes reception (telephone, telex, etc.), cleaning, rates, heating, building maintenance and insurance, security and parking. In the Venture Units, tenants are responsible for rates, building insurance, cleaning and internal maintenance. The service charge covers maintenance of the external areas and security (camera surveillance) of the grounds and external envelope.

At Warwick Science Park, Barclays Venture Centre leases are taken outside the Landlord and Tenant Act for periods of between one and three years and renewable. Start-up companies are offered a one-year term in the first instance. More established companies would be offered a two- or three-year term at a fixed rent. For the Advanced Technology Units leases are for twenty-one years and three days and on FRI terms. There are unconditional breaks at seven and fourteen years and conditional breaks by negotiation for tenants who move to other premises on the park. Rent reviews are at three-yearly intervals. Service charges for the Barclays Venture Centre units include the receptionist, heating of common areas, carpeting, water rates, security, insurance and maintenance of common areas.

Surrey Research Park leased premises are mostly on twenty-five-year FRI leases with five-yearly reviews. Some smaller units are on six-year leases with three-yearly reviews and the university takes on the responsibility of repairs and insurance which is paid through the service charge. Units in the Surrey Technology Centre are let on licences.

The service charge is broken down into two categories: an 'overall charge' for common parts and a 'development charge' for the parts not demised but not common to all. This charge varies from one building to another. For the six-year leases it includes external repair and maintenance. For the Surrey Technology Centre it includes the cost of a receptionist. None includes rates. The charge is based on costs plus a management fee. It is paid quarterly in advance and after auditing the charge is adjusted at the end of each year.

Belasis Hall Technology Park leases are available for six or more years, with three-yearly rent reviews for units over 2,500 sq. ft. (232 m^2). Leases are assignable with the landlord's consent. Three-year tenancy agreements capable of termination with three months' notice are available for units up to 2,500 sq. ft. Rents consist of a basic rent, maintenance rent and service rent. Maintenance rent covers external building maintenance, courtyard landscaping and hardscaping. It is calculated per square foot and is index-linked. The service rent is calculated on out-turn costs and reviewed each year. It includes insurance, refuse collection, street lighting, external window cleaning, estate security and maintenance of common areas. For the smaller units in Pavilion 1 it also includes heating and internal lighting, common areas, signing, reception facility and rates.

Structures and Style of Management

There is a wide variety of arrangements for the management of science parks. Although it is therefore difficult to generalise, three typical arrangements have emerged: the *ad hoc* informal team, the single on-site manager and the management company. These arrangements will be discussed first to show how science parks are managed and then to consider the issues involved.

Informal Teams

This is the cheapest and most flexible solution. It involves no full-time manager: the property management is done by a professional 'estates surveyor' from within the organisation which carried out the development whilst other aspects of management are left to the academic institution. This may seem a logical solution to the need for a variety of different professional skills and types of management. It is due much more, however, to the way in which each science park has evolved and the need to economise on management costs.

Informal teams of two or three people can be seen to work very well. Indeed, the people concerned in such teams often speak themselves against more formal structures, arguing that the informality enables them to react more quickly to tenant requirements. It is clear, however, that their success is dependent on both the commitment of the agency and on the motivation and priorities of the individuals, since the people concerned are managing aspects of the science park as one amongst other duties. When personnel change, the new incumbent's priorities may be different and the previous rapport and support lost.

The Cambridge Science park is managed jointly by the Senior Bursar of Trinity College and Bidwells, chartered surveyors. Neither has their permanent base on the science park. The college gives scientific advice as required in tenant selection. The agent is responsible for lettings and property management and acts as project co-ordinator in the development of the science park in general and the college's buildings in particular. The science park now takes most of two senior surveyors' time. In the early days it was just one of the surveyors' responsibilities as it still is of the college bursar. This arrangement, which works well, benefits from continuity of personnel and the understanding which has developed with the evolution of the science park.

Highfields Science Park, Nottingham, is managed directly by a surveyor from Nottingham City Council as part of wider responsibilities for development and management of the council's industrial property portfolio. The university is supportive but gives no priority to science park tenants. The university's industrial liaison officer (partly funded by the city council) is responsible for advising tenants with regard to facilities and personnel within the university and advises the city council over tenant selection. The science park is just one of his university duties. Nottingham City Council offers a business advisory service. A standing working party consisting of city council officers and university staff advises with regard to policy decisions. There is no on-site presence.

Listerhills management is split between Bradford University and English Estates. Tenant support and overall management is provided by the university's director of industrial liaison whose office is adjacent to the main university campus. He has no specific science park budget and is responsible directly to the vice-chancellor. They meet regularly with the regional manager of English Estates (Yorkshire and Humberside, based in Doncaster). Property management is carried out by English Estates. There is no formal agreement between the partners. This works well because of a positive approach and good personal relationships. Those concerned are able to react quickly and work towards common objectives. The city council is supportive of both short and longer term intentions. The intensive tenant support provided in the first year of operation (see above, p. 51) is no longer considered necessary.

The Single On-Site Manager

A more purposive solution is for an on-site commercial manager to be appointed or funded by one of the sponsoring organisations. This will be the initial contact person for day-to-day matters. Such a manager's main role is generally to identify and nurture links between the academic institution and the science park, to promote the science park and to foster the development of a community of interest between tenants.

The extent to which the commercial manager is concerned with property management will vary with the requirements for tenant support and the complexity of the lease negotiations. In most cases, however, the manager rarely has the time or expertise to carry out any property management functions beyond making introductions for prospective tenants and passing on complaints from existing tenants. Thus, as with the informal teams, property management remains the responsibility of the development agency or local authority which carried out the development.

This division of functions can work well provided there is long-term commitment to the science park objectives. Tenant selection, for example, is typically made by the estates surveyor in consultation with the on-site manager. If the two individuals co-operate there should be few conflicts. There are, unfortunately, cases of property managers' negotiating lettings without even informing the on-site manager. Other problems arise as a result, for example, of the property manager's unwillingness to be flexible enough to accommodate both the more discerning and the more impatient tenants. It is difficult to market science park premises as superior to the alternatives available if the developers are unprepared to finish units until they are spoken for. Equally, such an approach will lose the tenant who needs accommodation immediately.

Managers have often been appointed for their scientific and research experience, gained in a senior position in industry. Such a candidate may also have had commercial experience within industry. Occasionally, appointees have come with experience of managing directorships in industrial companies but without the research/academic background. Certainly experience and background are important and will influence the style of management: especially as there will probably be little assistance, perhaps only a secretary/receptionist. It is usually considered important that managers are scientists so that they can evaluate the potential of a possible new tenant and can hold their own in academic discussions. Some argue, however, that the promotional drive of an experienced businessman or woman is more important since the professional evaluation can be provided by departments within the academic institution. Undoubtedly, the positive style which such a figure would adopt could benefit science parks whose main aim is to attract and nurture new businesses for the local community. The disadvantage is that it can be difficult for a non-academic to build the necessary links with academics who tend to be suspicious and insular. It is probable that the manager who also has a research background will be more successful in persuading the university academic of the importance of technology transfer. The manager's industrial experience will both enable him to recognise possible applications and markets and to convince the academics that he knows what he is talking about. In addition, the manager may take on the task of providing business advice.

It is evident from the foregoing that on-site managers need a wide range of expertise as well as personal flair to combine insight into industrial, scientific and business requirements with an ability to manage people and resources in a sympathetic but stimulating manner. In addition, the manager will have responsibilities to the various partners involved in the science park: conflicting interests may be a problem.

Bolton Technology Exchange has an on-site commercial manager who is responsible for tenant management. He is also involved in marketing the scheme. This post is partly (55 per cent for three years) funded by the European Regional Development Fund, the rest being the responsibility of the college. The latter commitment has not been met fully and increasingly the borough has had to step in. Property management, including lettings, is carried out by English Estates. There is no regular liaison committee arrangement between the partners and the manager.

The Mountjoy Research Centre, University of Durham, is managed by an on-site commercial manager. His role is to ease the process of transferring scientific knowledge into the market-place by attracting suitable tenants, encouraging liaison with the university and dealing with day-to-day management problems. He chairs a monthly meeting with English Estates personnel and is a member of a strategy committee which meets quarterly. This is composed of senior members of the university and English Estates, together with an officer from the county council. His salary is paid in equal parts by the university and English Estates.

Management Companies

A number of science parks are run by management companies, usually a private company limited by guarantee and with no share capital, the equity and the guarantee being provided by one or more of the parent/sponsoring organisations involved in the partnership. Normally all the parent organisations (which may be a higher education institution, a local authority, a development agency, a financial institution or an industrial company) are represented on the board of directors. One or more full-time managers are appointed to run the science park: they are responsible to the board which typically meets quarterly. In some cases management committees may be formed. More often the executive staff are left to take day-to-day decisions and to report to the board as it has determined. Such management companies may have a team or a single professional employee: the more staff employed the more aspects of development and management with which they can deal.

Aston Science Park company, Birmingham Technology Ltd., was established in 1983 to enable the science park to be commercially orientated and independent. The board consists of members from Birmingham City Council, Aston University and Lloyds Bank, together with a managing director. There is a management team headed by the managing director and chief executive. It consists of: the director of finance, the director of business development, a bank manager (seconded) from Lloyds, one business development manager, a property development manager, two accounting staff, a marketing officer and seven secretarial/administrative staff. Consultants are also available to extend the capabilities of the management team which adminsters the science park and its own venture capital fund. It gives management and business advice together with technological assistance. It also runs a range of promotion support services for board meetings, conferences and exhibitions. It is thus involved in all aspects of managing a science park and the full range of services to tenants.

The University of Warwick Science Park Ltd. (UWSPL) is controlled by a board with equal representation from the four partners (see above, p. 38) and chaired by the vice-chancellor. A committee of officers of the partners meets in advance of the quarterly board meetings and provides professional guidance for the management team which consists of a director with responsibility for major projects and initiatives including attracting large companies on to the park and the development of new land, and a finance and administration manager with responsibility for day-to-day matters including property management, maintenance and marketing. They are supported by two secretarial staff.

UWSPL operates as a commercial business and without external financial support. Its major sources of income are recycled debenture interests, retained share of rents and services to tenants. During the first three years the energies of UWSPL were mainly directed towards property development and the more static type of services (receptionist, telephone, telex, desk-top publishing) provided in the Barclays Venture Centre. With maturity, UWSPL is beginning to direct its energies to the more dynamic services such as advice on taxation and VAT and joint marketing ventures.

Belasis Hall Technology Park (opened March 1988) is run by Belasis Hall Technology Park Ltd., a joint company in which ICI and English Estates have equal shareholding. The board, chaired by an ICI director, consists of three members from each partner. ICI has seconded a senior executive to be chief executive of the new company. His responsibilities include marketing the park, technology transfer and tenant services. Tenant selection for Belasis Court (as Phase 1 is called) and Phase 2 is a joint activity of the chief executive and English Estates.

Manchester Science Park Ltd. is controlled by a board consisting of twelve members, four each from the University of Manchester supported by UMIST, the city council and a group of four local sponsoring companies. The management team consists of a chief executive, supported by a secretary and caretaker, who is responsible for administration of the science park, advising the board on the formulation of policy, tenant negotiations, business advice to tenants, university linkage and marketing. For the second phase of development, a local firm of chartered surveyors will be jointly responsible for the lettings. This is considered desirable, in particular because longer leases will be involved.

Differences in management structures are the result of a number of factors: the evolution of the science park and its size, the commitment of and interrelationship between the partners and the priority given to the science park, compared with other initiatives. These factors will determine the resources available and whether the management is proactive or reactive in style. It is however evident that a formal, integrated management structure which sets out functions, priorities and powers will be a more secure basis for longer term management and development. In this context, full-time, on-site management is highly desirable. Although constraints on resources may make this seem an extravagant expense when there are few tenants, the evidence shows that it helps speed up the initial growth rate and brings forward the date at which the science park can be considered to have acquired its own identity and momentum.

References

1. Scarrett, D., *Property Management*, Spon, 1983, p. 3.
2. Rowe, D., 'Achievement from enterprise', UKSPA Conference, London, February 1988.
3. Storey, D. J., et al., 'The high tech entrepreneur and his business', UKSPA

Conference, London, December 1986, Table 11. See also below, Chapter 6, n. 6.

4. Monck, C. S. P., et al., *Science Parks and the Growth of High Technology Firms*, Croom Helm, 1988, p. 30.
5. Venture Economics Ltd., *Survey of the Impact of Venture Capital in the UK*, British Venture Capital Association, March 1987, p. 2.
6. Venture Economics Ltd., *Report on Investment Activity 1986*, British Venture Capital Association, 1987, p. 4.

6. TENANTS OF SCIENCE PARKS

Characteristics of Tenants

Over half the tenants on science parks (Figs. 1 and 2) are less than three years old and almost half employ no more than five people (Figs. 3 and 4). The majority of tenants are single plant independent companies (Fig. 5). A comparative study[1] based on a sample survey showed a situation on conventional estates and high technology developments very different from that on science parks: the percentage of independent companies was 27 per cent, 37 per cent and 63 per cent respectively.

Fig. 1 Total number of tenants on UK science parks (*Source, UKSPA*)

Year (end)	*No. of tenants*
1985	288
1986	412
1987	641

Fig. 2 Employment growth in UK science parks (*Source, UKSPA*)

Year (end)	*Total no. of employees*
1985	3,469
1986	5,311
1987	7,632

The age structure and size of firms have changed as the new firms established on the early science parks have matured and as the parks have attracted a wider range of tenants (Figs. 3 and 4). Most firms are in the well-recognised high technology sectors of computing, electrical engineering, chemical and biotechnology and consultancy (Fig. 6). The study showed a higher level of sophistication on science parks than on other sites. The proportion of occupiers classified as high technology industry were: science parks 68 per cent, high technology developments 44 per cent, and conventional estates 24 per cent.[2] Ninety per cent of science park firms in a recent national survey were doing some research and development, one third was engaged in 'radical' new research and two thirds in developing new or complementary product ranges.[3] Employees of science park firms are also significantly better educated than non-science park firms.[4]

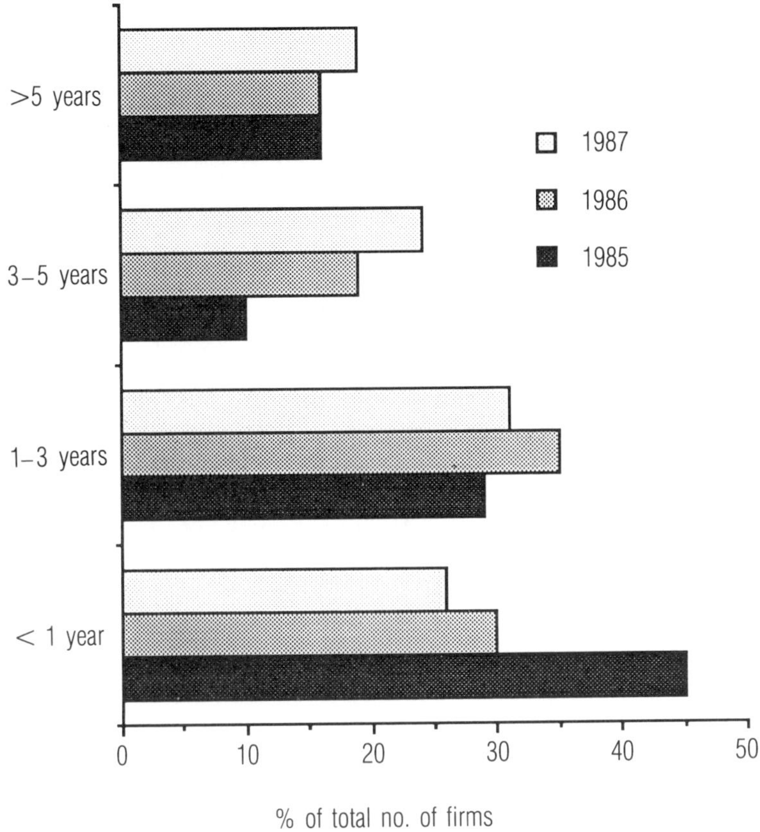

Fig. 3 Age of firms on UK science parks (*Source, UKSPA*)

Tenants' use of floor space on science parks shows broad similarities with those on high technology developments.[5] The main differences are more laboratory space on science parks and more storage on high technology developments. Tenants on conventional estates show a markedly different usage (Fig. 7, see p. 68). Proportions will vary with individual tenants and the distinctions for example between office and research and development will be more or less clear depending on the industrial sector and the organisation of each firm.

Growth

The growth rate of small firms on science parks appears to be far in excess of high technology firms located elsewhere. Research involving interviews with a sample of firms carried out on behalf of UKSPA[6]

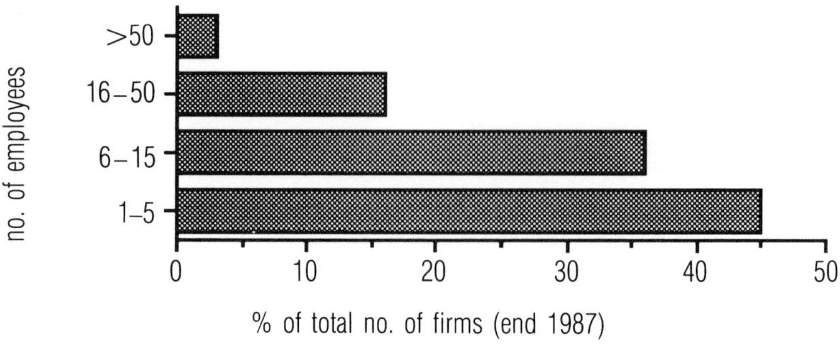

Fig. 4 Size of firms on UK science parks (*Source, UKSPA*)

during 1986 showed that employment increased between 1985 and 1986 by 80 per cent among science park firms, 28 per cent amongst high technology firms not on science parks, whilst small firms generally showed a one per cent decrease.[7] Also science park firms have a very low failure rate (see above, pp. 49–50) and a greater likelihood of requiring expansion space.

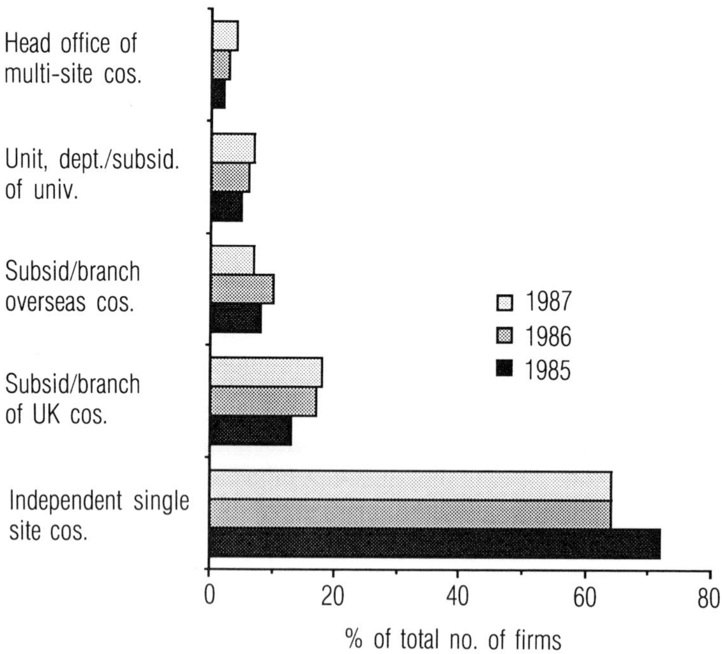

Fig. 5 Status of tenants on UK science parks (*Source, UKSPA*)

Fig. 6 Industrial sectors of firms on UK science parks (*Source, UKSPA*)

Industrial Sector	*% of Total No. of Firms* 1985	1986	1987
Computing	33	34	34
Hardware and systems	16	11	8
Software	17	23	26
Electrical	18	19	17
Microelectronics	8	4	4
Instrumentation	5	5	5
Automation	2	6	5
Electrical equipment	3	4	3
Chemical and biotechnology	10	12	15
Medical	5	5	6
Pharmaceutical	3	5	6
Fine chemicals	2	2	3
Consultancy	16	19	19
Design and development	4	4	4
Analytical and testing	3	4	3
Other technical services	9	11	12
Financial and business services	8	8	6
Other	15	7	9

Individual firms frequently show exceptionally volatile growth characteristics. A firm may be satisfied with its small unit one day but in a matter of months want to double or quadruple the amount of floor space. Another may sign an agreement to lease 5,000 sq. ft. (465 m^2), yet, whilst the building is under construction, need to double its floor space. Firms undergoing such rapid growth are typically not conscious of their increased accommodation requirements until it is actually needed.

Small innovative firms often display a cyclical nature in their activities. These and consequently their use of floor space, will vary according to the stage of development of a product. When a new idea is being developed, laboratory/workshop space for research and development will be major needs. When a more standardised product emerges, manufacturing and sales will be more significant. When sales drop or further ideas evolve, research and development will again take precedence, the product will be redesigned or replaced. The time period of this innovation–production cycle will vary and some products will never reach mass production. A five-year cycle, however, appears to be common in small high technology firms.[8] Investment needs and profit will also change as the cycle progresses. Figure 8 shows investment and production cycles

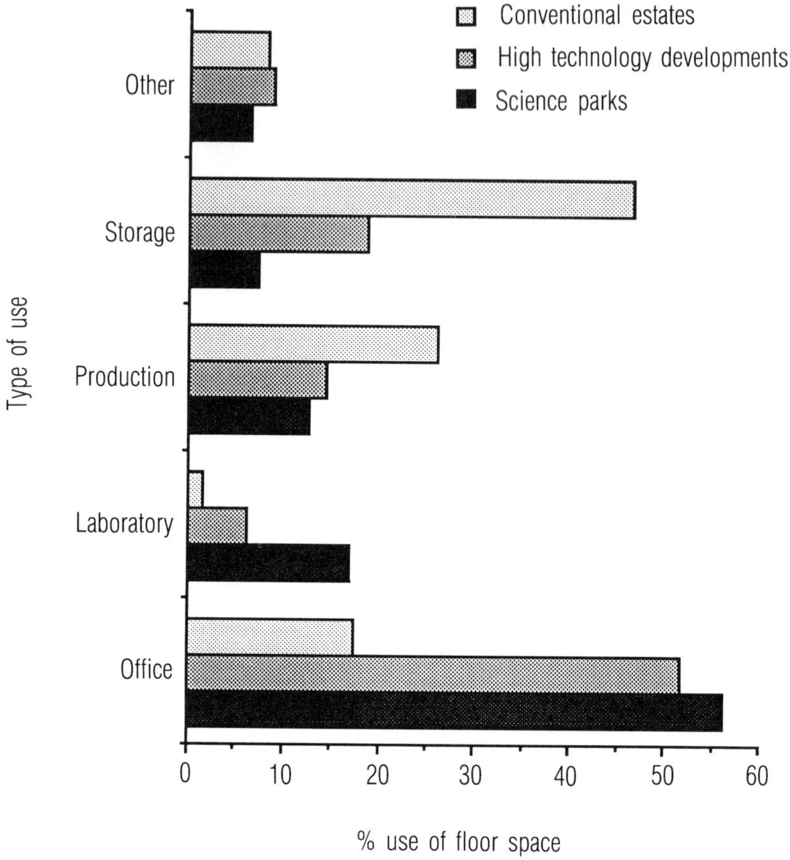

Fig. 7 Tenants' use of floor space, 1986 (*Source, Henneberry, J.*[5])

in a single-product firm. Such firms tend to be subject to uneven financial pressures. Those firms with multiple product lines will clearly be more robust.

Origin

In terms of origin most science park firms come from within the same travel-to-work area. Few have come from another region of the UK and very few from overseas. A survey by UKSPA, in November 1985,[9] showed 17 per cent had come from the parent higher education institution and another 66 per cent from within a five-mile radius. Only 14 per cent were from places in the UK over thirty miles distant and 5 per cent from overseas. This survey also showed that 42 per cent of the total

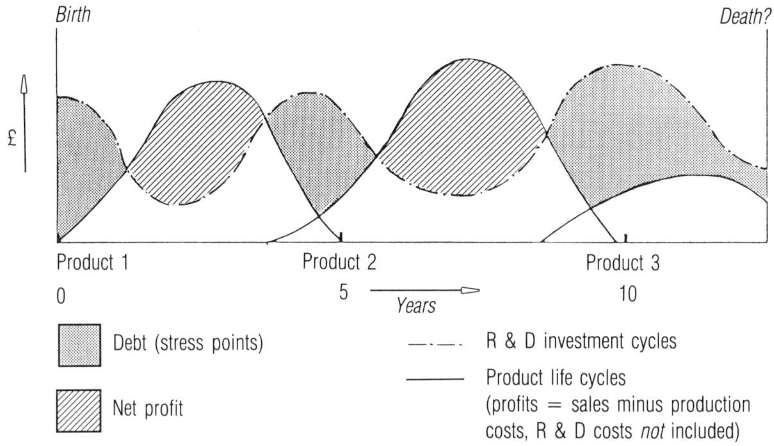

Fig. 8 Research & Development investment and product life cycles in the small high technology firm (*Source, Oakey, R.*[8])

were start-up companies. There is considerable variation between sites. In October 1987, well over half of the firms at South Bank Technopark were new companies, at Warwick Science Park only a quarter.

If firms continue to be mainly local, there are implications for the size and number of science parks. Some schemes have, however, even initially managed to attract small and medium-size firms from other regions of Britain. At Warwick, half the firms are part of corporate UK ventures and almost a quarter part of overseas companies. Durham has also been able to attract a substantial proportion of firms from other regions of the UK. Once a science park site is established it will look more attractive to mature companies. It is likely that science parks will be competitors in the search for these tenants.

Funding

Obtaining funding is a major problem for start-up companies, particularly for a firm involved in research and development which may need a long lead time before a saleable product is available. The UKSPA survey found that 33 per cent of firms in their first year of operation considered this a serious obstacle.[10] Banks look for stability while venture capital funds find it more profitable to lend larger sums than new firms either need or can afford (see above, pp. 51–52). Personal savings are the most important source of start-up capital (55 per cent of tenants in the UKSPA survey),[11] clearing banks provided capital to 12 per cent; other less important sources were mortgages, previous businesses, government grants. Only 3 per cent were financed by venture capital. As well as the general reluctance on the part of venture capital companies to provide seed capital, a further problem exists for firms north of a line from Bristol to Cambridge. Although they show no significant difference in terms of research and development effort or in the technological novelty of their product, there are clear differences in both the level of external equity participation and business growth and ambition. The same survey[12] showed that venture capital participation, employment growth and ambitions for USM listing were twice as high in the south as in the north. In 1986 venture capital investment in the south-east and Greater London was 64 per cent of the UK total.[13]

The emergence of the 'hard' and 'soft' company models (see below, p. 74) has significance in terms of funding. The hard model depends on external capital sources and thus the availability of venture capital will affect the growth prospects of such firms. The soft model is less dependent on venture capital in the early years. It is likely to have lower fixed costs and long-term contracts with particular clients. Both government departments, especially the Ministry of Defence, and large companies may contract such small specialised firms as suppliers.

Linkage

Academic linkage

Interaction with universities, other higher education or research institutions can take many different forms: casual to contractual, social to technical. To dismiss linkage between a science park and its parent university by saying that very few companies were set up by university academics is to over-simplify the issue to the point of misunderstanding. Spin-outs from the university are only one of many forms of linkage.

Science park firms put access to the higher education institution high on their list of reasons for locating on a science park.[14] The UKSPA survey showed that 70 per cent of science park firms have informal contacts with academics; 36 per cent employed staff from the higher education institution on a consultancy or part-time basis; 40 per cent of firms on science parks use specialist equipment from that institution; and one-fifth sell goods or services to the institution. The fact that firms locate on a science park to be close to a higher education institution does not mean that linkages actually exist: a survey carried out for English Estates at Listerhills in 1987 showed that firms have been content that their image benefits from a university association and, consequently, there has been little interest from firms in drawing in any formal sense on the expertise and resources of the university. Warwick University, on the other hand, is able to point to several examples of working linkages. Even those firms only hoping to benefit from a good address should not be dismissed: it shows an appreciation of the importance of contacts and, indeed, such firms have often developed real technological linkages after moving in. Access to and use of other academic or research institutions is also important. The UKSPA survey showed twenty per cent of firms making use of other locally based research institutions. A series of surveys at the Cambridge Science Park in 1983, 1985 and 1987 also shows local research linkages other than with Trinity College to be significant.[15]

The Surrey Research Park has examples of various forms of linkage. Here, tenant companies must be engaged in research and development (which is defined very broadly) but not in manufacturing. The lettings policy does not, however, stipulate linkage with Surrey University: it looks at the type of activity in which the firm is engaged and its employment profile. It is interesting, therefore, to note that actual linkage with the university includes: joint short courses, joint research contracts, joint appointments, consultancy agreements, one-off research contracts and use of university equipment. An example of the latter is use of the university's main frame computer. There is provision for all tenants to link in; about half have taken up the opportunity. Linkage should be looked upon as a network of opportunities which companies may take up as needs arise: the richer that network, the more likely that there will be demands for access to it; that it will be a primary location factor for prospective tenants and that real benefits to industry or academia will accrue.

Cambridge is interesting in this respect. An exceptional network exists. As well as a university of thirty-one colleges with an enormous output of natural science and engineering graduates, a large number of government and other research establishments are represented in and around Cambridge.[16] The university attracts both students and academic staff who combine competence and self-confidence and whose attitudes are critical. In all academic, professional and business circles people express satisfaction with Cambridge as a place to be, for living and working. A not uncommon reason for starting a new business there is a desire to stay in the town. Interaction within the network has been stimulated by a unique combination of factors: the non-centralised collegiate structure of the university where social and academic affairs intermingle – no nine to five university here; a long tradition of interaction between academics, industrialists and businessmen and the fostering by Barclays Bank of a link through the establishment of the Cambridge Computer Group in 1979.

Belasis Hall Technology Park offers access to the whole range of services, facilities and resources of ICI (Teesside), where there are over 3,000 technical staff, of whom some 1,550 are post-graduates or post-doctorates (i.e. more graduates than most universities). Small firms on Belasis are thus able to benefit from expertise and use of equipment and services which normally only large companies can afford. Much of the advice will be free, equipment will be charged for at cost on a time-share basis. Furthermore, whereas universities are particularly able to help at the research and development end of the innovation-production cycle, ICI will be able to offer expertise in production, sales and marketing as well. As yet it is too early to assess actual linkage; it is already clear that potential linkage is a real attraction to small and new firms. Medium-size firms will perhaps hope to benefit more from a mutual association. Indeed demand was so strong that the first 50,000 sq. ft. (4,645 m^2) of small and medium-size units were virtually pre-let and by the end of 1988 their firms were in occupation. A further 55,000 sq. ft. (5,110 m^2) are under construction and another 100,000 sq. ft. (9,290 m^2) were about to commence in early 1989 (see above, p 34). Belasis Hall is also developing linkage with Durham University through the Mountjoy Research Centre (a distance of twenty miles). The latter, which wishes to concentrate on research and development, sees Belasis Hall as a possible location for firms moving into manufacturing.

Industrial Linkage

Interaction with industry is at least as important as academic linkage. Many science park managers consider the creation of a community of interest between their tenants as an important objective. Many tenants wish to be part of a community of like-minded firms: they benefit in terms of recruitment, exchange of information and ideas as well as social contacts. Small and even medium-size firms benefit also from sharing equipment and joint working on the same contract. Being within close range of particular firms is often cited by science park tenants as a reason for being there. Furthermore, if a community of interest is built up, this is likely to be a determining factor in whether a tenant moves elsewhere or stays within the same site when it wishes to expand. Science park managers will recognise the importance of holding tenants with growth potential.

At Bolton Technology Exchange a consortium of all the firms on site, together with two large off-site firms, has emerged. Individual skills can be combined to work on single projects which are managed by the commercial manager. Access is available to relevant expertise and other resources in the Bolton Institute of Higher Education. Since 1987 they have been tendering for contracts in advanced manufacturing technologies. This arrangement encourages technology transfer whilst enabling small firms to handle larger contracts than they otherwise would and thus stimulating their growth.

Wavertree Technology Park illustrates the way in which a mature industrial company may take a lead role comparable to that of an academic institution. Plessey is one of the partners in the Wavertree Technology Park Company (together with Liverpool City Council and the former Merseyside County Council). In addition to seconding staff to the WTPC management team, Plessey makes business and technical expertise and consultancy, design and other services available at cost to tenants of the park. The WTPC has also developed links with Liverpool Polytechnic and the University of Liverpool. (See also in this context, Belasis Hall, p. 71.)

New Firm Formation

The nurturing of firms which spin out from the parent academic institution is seen as a *raison d'être* of science parks. 'Spin-outs' are the most visible sign of technology transfer, while the accommodation needs of such companies have often been an important catalyst in the process of a university setting up a science park. The case of Metals Research, which was forced to move to a site nine miles out of Cambridge in 1966, caused sufficient consternation in the university to trigger active moves to ensure that it could not happen again. Spin-out firms have also been persuaded to be the first tenant on a site in an attempt by management to give credibility to a new development (e.g. Laser-Scan at the Cambridge Science Park and Bradford University Software Services at Listerhills, Bradford). Local industrial companies which have substantial in-house research, large consultancy firms or research institutions are other sources of spin-outs. A sign that a science park has achieved a momentum of its own is when spin-outs from firms on the same site take place. It is worth noting that eighty per cent of new high technology firms originate from industry and only twenty per cent from academia.

Favourable Conditions

The conditions which will be favourable to spin-outs are of interest to developers and managers of science parks. The factors which will motivate an individual to set up a new business are complex: some have a 'push' effect causing people to want to leave, and others a 'pull' effect because people see opportunities. In this context the nature of academic employment conditions (contractual and others) and the types of people in academic posts are important. The British national character, often put forward as a factor inhibiting entrepreneurial activities, especially in comparison with the USA, can be dismissed for our purposes: our concern is with the significant differences which occur within the UK. Cambridge, where large numbers of spin-outs have occurred, has been studied in an attempt to identify the factors in this so-called Cambridge phenomenon[17] with a view perhaps to repeating it elsewhere. Conclusions have emphasised the uniqueness of Cambridge. These studies, together with comparisons at other science parks, do, however, give some useful indicators.

Firstly, the presence of a large organisation with substantial leading-edge research and development activity is not in itself enough for spin-outs to occur. There are many factors which act as real 'barriers' to staff exit. It may also be necessary for there to be genuine insecurity factors such as the threat of unemployment. Compare, for example, British Nuclear Fuels Ltd. at Warrington and British Aerospace at Bolton, where no spin-outs have occurred, with ICI at Billingham. The cut-backs in staff at all levels in Billingham are more likely to generate tenants for Belasis Hall Technology Park. In Cambridge, frequent staff exits from some firms owing to bad management was undoubtedly an ingredient in the Cambridge phenomenon.

Secondly, the nature of the 'barriers' varies from place to place. The flexibility of contracts of employment at Cambridge University and the university's attitude to patents certainly acts as an enabling factor. This low threshold environment allows academics 'to test the water' over a period of years. It may be, however, that a major barrier is the academics themselves. They have chosen a university career because of the intrinsic interest of the job not for the money. Thus they find little attraction in making extra money. In addition, promotion depends on academic performance, running businesses does not count. Contract research is seen as a dilution of fundamental research and therefore to be avoided if possible. Financial cut-backs in higher education may increasingly, however, push academics into 'testing the water' and perhaps even change the criteria for promotion.

The process of 'hardening' of new firms is neither simple nor linear. Bullock[18] identified two models of new firm formation: the soft and the hard models. The 'soft company' model emerges gradually from academic institutions with a low threshold environment. They follow a distinctive pattern, starting as technical consultancies and passing through stages of routinisation of services and products, eventually to become manufacturers of standardised products. Thus the 'soft' company undergoes a process of 'hardening' from highly analysed individual solutions to problems to standardised responses. The 'hard company' model is different. Here the inventor 'spins out' of the parent environment setting out to produce a standardised product almost from inception. The soft model is less risky, being easier to set up and requiring lower levels of finance and management in the earlier stages. These models are undoubtedly useful in understanding the process of new firm formation. More recent work in the UK[19] shows that firms may move from 'hard' to 'soft' as well as from 'soft' to 'hard'. In reality, the process of growth of firms may be much more complicated.

Finally, the quality of the 'community of scholars' which exists in and around a university is important. It needs to be nurtured if it does not exist, a fact recognised for example by Aston, Surrey and Warwick. In many red brick universities, academic life has meant isolation from the local community: an academic's job has required little contact with employees elsewhere and home has been miles away in a different community, often owing to the lack of good housing close to the university.

Although the growth prospects of their tenants make science parks an attractive opportunity for investment, the nature of such firms indicates a need for developers and managers to have a greater understanding of the tenant than would normally be the case. The factors which encourage new firms to emerge and grow, the process of their evolution and their specific property and other requirements are necessary elements in this understanding. The origin and status of science park firms clearly have implications for any development appraisal and the subsequent marketing of a scheme.

An important aspect of a mature and successful science park is a range of tenants of different sizes and with complementary technologies which have a high level of interaction. Those tenants will require a range of different types of premises and leases. It is thus in the long-term interests of a science park to nurture new firms and to be able to provide those firms with new or enlarged premises just at the right time to suit their own business.

References

1. Henneberry, J., 'High technology occupiers and their use of accommodation', *Land Development Studies*, 4, 1987, p. 127.
2. Henneberry, J., ibid, 1987, p. 132.
3. Weild, D., & Quintas, P., 'The technological relevance of science parks', UKSPA Annual Conference, London, December 1986.
4. Weild, D., & Quintas, P., ibid, 1986; Carter, N., & Watts, C., *Cambridge Science Park*, Planning and Development Case Study 4, Surveyors Publications, 1984, p. 38.
5. Henneberry, J., op. cit., 1987, p. 138.
6. The UKSPA survey covered 149 science park tenants. Cambridge, Aston and Heriot-Watt were excluded: Cambridge and Aston had been used for the pilot survey and Heriot-Watt was omitted as the other first generation science park. The survey is thus considered to be a review of more recent or second generation UK science parks (Storey, D. J. et al., 'The high tech entrepreneur and his business', UKSPA Conference, London, December 1986).
7. Monck, C., 'Science park tenants and their growth potential', UKSPA Conference, London, December 1986.
8. Oakey, R., *High Technology Small Firms*, Frances Pinter, 1984, pp. 92. and 94.
9. Monck, C., 'Growth of science parks – a progress report', *Science Parks – their Contribution to Economic Growth*, Proceedings of the UKSPA Annual Conference, December 1985, UKSPA/Peat Marwick, 1986, p. 8.
10. Storey, D. J. et al., op. cit., December 1986, Table 11.

11. Porter, R. B., 'Management and development issues for science park companies', UKSPA Annual Conference, London, December 1986.
12. Storey, D. J. et al., op. cit., December 1986, Table 12.
13. Venture Economists Ltd., *Report on Investment Activity* 1986, British Venture Capital Association, 1987, p. 8.
14. This is shown by the UKSPA/Peat Marwick McLintock survey 1986, and by the author's own research over several years, covering most UK science parks.
15. Surveys carried out by the author of all tenants of the Cambridge Science Park.
16. Carter, N., & Watts, C., op. cit., 1984, pp. 6-7; Segal Quince and Partners, *The Cambridge Phenomenon*, Segal Quince and Partners and Brand Brothers, 1985.
17. Segal Quince and Partners, ibid, 1985.
18. Bullock, M., *Academic Enterprise, Industrial Innovation and the Development of High Technology Financing in the United States*, Brand Brothers, 1983.
19. Weild, D., & Quintas, P., op. cit., 1986.

7. CONCLUSION

Science parks have secured an identifiable place in the spectrum of planned employment environments. At the same time they are but one of many initiatives (mostly non-property based) intended to encourage technology transfer and stimulate job and wealth creation. Property alone is not enough to ensure the achievement of the non-property objectives which are such an essential part of science parks, and which distinguish them from business parks.

Space in a science park is neither a sufficient nor a necessary reason for the growth of high technology firms. Many such firms thrive on technology parks or business parks where the emphasis of developers and managers is primarily to maximise financial investment. Furthermore, in the more popular areas of the country for the growth of new industries it may not be necessary to sacrifice shorter term financial profit to achieve the longer term economic goals. Thus science parks do not all strictly meet the definitions given at the beginning of the book in terms of priorities. Much depends on the status and resources of the developer as well as the objectives of the individual science park. Some science parks quite clearly give property objectives a high priority. Nevertheless there is a science park category of development, a distinctive sub-sector of the industrial property market, which has proved its growth potential and which offers enormous scope to the property world.

Science parks have reached a critical stage of evolution in that future directions depend on the outcome of economic, political and financial decisions by the agencies involved and affected. The various objectives frequently conflict and decisions may be mutually exclusive. There is a role for stronger partnerships between interested parties and for a clearer lead from government.

There are three broad issues for those involved in property development and management of science parks.

Making the Best Use of Science Parks

Benefits can be seen in terms of the multiplier effect in the local economy and the dialogues initiated. Employment growth is impressive but in itself makes little impact on unemployment totals. The significant factor is the extent to which a science park can act as a focus and a catalyst for growth by stimulating change within a given local economy. The process of developing and managing science parks has initiated dialogue between the participating bodies, the benefits of which go far beyond the science park itself. Dialogues, between local authorities and higher education institutions, between higher education and industry, between industry and local authorities and between higher education and the financial institutions now exist where previously there were none. A climate of confidence and mutual respect has been created.

Success so far, however, is limited: partly because of the extreme youth of most schemes but also because of the difficulties of harnessing academic ideas and inventions which have industrial potential. The low level of interchange is frequently blamed on academia, but industry itself should also be criticised. In general, UK industry does not respect academia and does not know how to use the talent fostered in the academic environment.

The attitudes and roles of the major parties involved is a critical determinant of the success of any science park venture. If the motive for establishing a science park is merely to keep up with the herd and the behaviour of the sponsors and managers reactive and negative rather than positive, the venture is unlikely to be worthwhile. The attitude of local authorities to the higher education institutions in their locality is frequently almost resentful. Those local authorities which have seen their local university as a positive asset and acted accordingly have certainly benefited.

The science park will require a continuing commitment in financial as well as in management and general support terms. The venture needs to be seen as a long-term project which will continue to generate benefits to both the sponsoring agencies and the local economy.

The role of central government should also be questioned. To date, it has been very passive compared with Japan and other Western European countries. To some extent it helps by creating an environment for enterprise and giving very limited financial support to innovative ideas. It could help also by a more positive and co-ordinated policy at both regional and national levels. The crucial role of new high technology firms is not fully appreciated by those who make science policy. The role of the UKSPA is relevant in this context. Some members feel that it should take on a more active approach to lobbying government and financial institutions. The latter will need to be involved more if science parks are to achieve their longer term objectives.

Achieving Success

Successful science parks have a number of characteristics in common. A high standard of buildings and environment is a prerequisite; they need also to be managed in such a way as to maintain that standard. The continued commitment of the higher education institution (or other research centre) and support by other local bodies concerned with research, industry, finance and professional services will be vital. A sustained high level of general promotion and public relations to stimulate awareness within the academic centre, the local community and in a regional and national context will be important, as will selected and targeted marketing. Advice and support for tenants, especially young firms, will pay off in terms of low failure rates and rapid growth. Fostering linkages of all kinds will assist firms to grow and make it more probable that they will wish to stay within the same site or local area. Thus, resources will need to be devoted to all aspects of management if the property initiative is to be justified.

Multiple sponsorship of science parks, including private and public sectors, should be more readily able to sustain the continuing management required, both in terms of resources (financial and other) and commitments. Formally structured partnership arrangements are likely to lead to more lasting success and to be able to attract investment for further growth. The most difficult period will be the first year or so. Once the park is operating and has a clearly visible presence it will be easier: nevertheless a positive approach will be needed to keep the momentum going. Therefore, good management and a close relationship between management and development and between management's objectives and the science park's objectives will be essential.

As well as commitment and management, success requires resources: technological, professional and financial. The latter are required in the form of funding for further property development and risk capital for tenant firms. In both of these areas, shortage of investment has inhibited growth. The prospects for attracting appropriate investment are, however, increasingly optimistic.

Investment by the private sector has gathered momentum during 1988. 3i is exploring other science park development opportunities, and recent schemes at Warwick and Silwood are expected to be attractive investments for financial institutions. The privately funded business centre at Belasis Hall shows that development companies do see benefits in developing small units and providing back-up services. The rapid pace of development there has encouraged BHTPL, within nine months of opening, to consider funding the next fifty acres of infrastructure itself. Furthermore, Science Cities Ltd. is aiming to be the most comprehensive venture yet. This is a private development company set up by four people, with considerable science park experience, to provide specialised development and management of science parks, including on-site management and training courses. Completed schemes will be able to take advantage of a network linking them with other Science Cities schemes. Three are agreed: Newcastle (involving both university and polytechnic), Oxford (Magdalen College) and Ashford, Kent (Trinity College, Cambridge). Several others are under negotiation. In addition, the company hopes to take over the management of some existing parks.

The Future

The main impact of most existing science parks is in the local travel-to-work area, the local economy and the parent academic institution. Many could enhance that position by adopting a higher profile and building more linkages. Further expansion must be built on that local base. For the most part, science parks have not been in competition with each other. As they expand, however, they will need to attract larger firms, many of which will come from other localities. The extent to which large firms need a science park will also need to be examined. Science parks may be increasingly in competition with each other for funds and firms. The availability of land and planning policies may well have restraining effects.

Schemes are thus likely to become more divergent and to involve a wider range of partners, especially from the private sector. Those higher education institutions or local authorities which open a science park in imitation of their fellows should think again about the purpose of the venture and the means of achieving it. Undoubtedly science parks will continue to be a growth area in property development and to attract more commercially minded partners. Initial feasibility studies, thorough evaluation and effective implementation will be essential ingredients of successful ventures.

INDEX